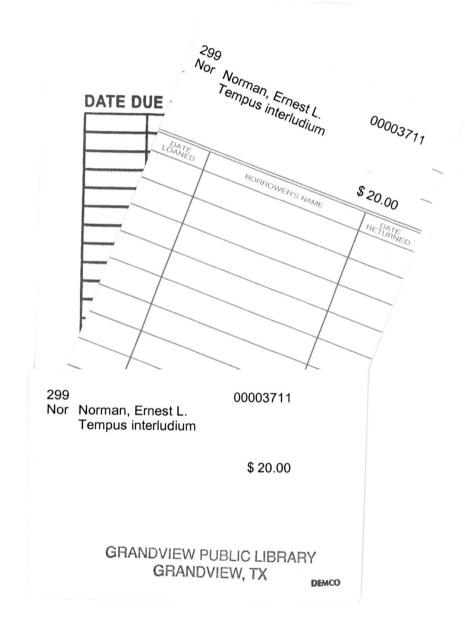

DATE DUE

299
Nor Norman, Ernest L.
Tempus interludium

00003711

DATE LOANED

BORROWER'S NAME

$ 20.00

DATE RETURNED

299
Nor Norman, Ernest L.
Tempus interludium

00003711

$ 20.00

GRANDVIEW PUBLIC LIBRARY
GRANDVIEW, TX

DEMCO

Ernest L. Norman
Author, philosopher, poet, scientist, director,
moderator of Unarius Science of Life

UNARIUS
UNiversal ARticulate Interdimensional

Understanding of Science

TEMPUS INTERLUDIUM

VOLUME TWO

TEMPUS INTERLUDIUM

PART TWO

BY
ERNEST L. NORMAN

UNARIUS
EDUCATIONAL FOUNDATION
145 S. Magnolia Avenue
El Cajon, California 92020

ISBN 0-932642-48-9

UNARIUS EDUCATIONAL FOUNDATION
EL CAJON, CALIFORNIA

CONTENTS

PART TWO
DISSERTATIONS FROM VARIOUS BROTHERS
AND
TEACHINGS OF THE MODERATOR

Note

The articles contained in part 2 of this volume titled "Tempus Interludium" were dictated by the Moderator, Ernest L. Norman, throughout the years in response to current news as it was related via television, radio or the news press. Because the earthman views all things from the third dimension only, very often the Moderator would be prompted and inspired immediately upon hearing of these many and varied reports, and would begin a dissertation without a moment's notice! Very often I would have to scamper quickly for a pen and pad as He did not favor the recorders; rather, He preferred the human element to aid him.

And so day by day, month after month, were these current issues responded to and dealt with from his Higher Self. I had more or less set them aside for the day when there would be the time and cycle to publish his vast accumulation of these, his over-the-years voicings. Now is the appropriate time and cycle for these invaluable words from our great and beloved Moderator to be made available for the public.

Bear in mind, these articles were not dictated in any one or few sessions but rather, over the many years time, and in direct response to current news events were they told to me (Ruth, the wife of Ernest) as He listened to the struggling earthman always striving from his reactionary ways, his negative, karmic past. But this he must do; must do until he begins to conceive the wondrous new way of life as brought to Earth by the countless Higher Spiritual Minds, the Leader of whom is the Unarius Moderator, Ernest L. Norman, whose teachings shall become the New World Science.

— Uriel

PART ONE
LETTERS FROM THE MODERATOR IN
REPLY TO QUESTIONS FROM STUDENTS

Letters to Students
From the Unarius Moderator

Dear Unariun: - It is always a happy and joyous occasion when we can add another name to the roster of Unarius, for we know, not only through our own personal experiences, but also from many others, that success has been attained in their efforts to renovate their lives. To most students who first contact Unarius, there is an immediate and tremendous quickening of the Spirit, and all of the pent-up emotions, complexes, and inhibitions are poured out like an onrushing tide, demanding immediate attention; and so he thinks that all of the things which he has collected in a thousand or more lifetimes must be justified, all of the hidden mysteries must be revealed, his body healed and his house set in order in just a few short days.

The true meaning of life is one in which every individual attains spiritual solvency by attaining the wisdom of self-mastery through a countless and infinite number of lifetimes and their experience quotients. Likewise, the horizons of any individual's personal perspectives are not broadened by having someone else elevate us into a higher plane of consciousness. Wisdom is a personal integration of an unknown and infinite number of factors. Through the lesson course, the most important and initial steps have been taken whereby you, too, can climb the same stairway used by all great Masters and Avatars. The most formidable of all diseases is ignorance, a worthy adversary, overcome only by the sharp edge of wisdom, and a part of the armor worn by you to attain mastery over ignorance will be your conceiving the principles.

1

All things in the eye of the Creator move forward in perfect order and harmony, an expanding and contracting Universe, linked and relinked through an endless succession of cycles. Intelligence from the Great Mind is manifested through an infinite number of dimensions, and all such wisdom becomes common property to those who use and understand the laws which govern cycular patterns in their linkage and in their proper timing through frequency relationship. As part of the all-creative Universal Cosmogony, you, as an individual, are thus subject to the same interpolations as are manifested in all creative processes. Proper timing, therefore, is the essence, and in using the word time, this is a direct reference not to the round face with the two hands and the figures, but to cycular patterns, their movements, and proper junctions.

Unarius is an exact science, and to take full advantage of its many ramifications, you must first learn to live through the directorship of the Superconscious, to rely upon the infinite power and wisdom contained therein, and to cease being the reactionary creature of the past ages.

The longest journey begins with the first step, and the first step of your long journey into Eternity will be to stop trying to direct your destiny with your conscious mind and its affiliations with the subconscious. Your self becomes the most important moving force which activates your own personal cycular pattern. In other words, by proper application in studying the lessons and other factors which have become adjunctive since your enrollment, your present mental position will be moved forward to the point where you can gain the greatest value from your reading. Be patient, therefore, for impatience can, in itself, generate an energy-block which will make it impossible for you to receive all help directed to you until such blockage is removed.

It is therefore very obvious and extremely important that you do your part by maintaining as clear and as untroubled a position as is possible under your present circumstance. Also remember, that after you have discharged the impeding blocks of psychic shock in the course of your therapy, there is yet a great deal to be done.

Like the slum clearance programs carried out in large cities, you too, must begin to clear out the slum sections in your own mind which are the old reactionary thought patterns which have been built up around your old conditions and your reactionary way of life. Jesus' words, "Go your way and sin no more," are a direct reference to stop using these old insidious and poisonous thought patterns which are always motivated by old sins and evils, because these same thought patterns will bear the progeny of new evils. Rest assured, therefore, that since it is in the Mind of the Creator that all things are whole and perfect, and since into the eternity which is ahead of you and its countless number of lifetimes, you will learn to justify through faith, wisdom and application, this same perfection. Fear not, for likewise Godliness shall be added unto you.

Meanwhile, during the course of your therapy, in the immediate and near future, the Radiant Energies shall be projected your way and all healing agencies constantly directed unto you, as of and until such time when you have learned to walk in the Light cast by the Infinite Mind.

Dear Student:

Following are a few vital points which will help you to better understand the corrective processes which enter into spiritual healing. The principles and methods are largely contained in the lessons. You will find by careful study, a scientific presentation of mass and energy and that all things resolve into energy. Therefore, in the creation of man, there is a broad departure from the Biblical, physical version of the creation of Adam and Eve.

Continuing along this line of thought and pursuing our course of reasoning, we therefore resolve man in all the tangible and intangible elements of God's Infinite Nature. Man is therefore, individually and collectively, constantly revolving through countless dimensions, lifetimes and experiences. This is motivated not from the physical plane of existence or expression, which is only effect, but from the spiritual side of man's true self.

It is well-known that man has a spiritual body, but very few people seem to know what that spiritual body is. I do more than presume to know, and can factually prove that it is an energy body existing in another dimension—outside the third-dimensional concept of time and space. This energy body is composed of the sum total of many lifetimes and evolutions. It also has a higher body, or the Oversoul, or the Christ Self, which is the motivating life force and which is made up from the abstract Infinity of God, Himself.

Experience is implanted or reflected into this spiritual or psychic body as definite wave forms of energy, residing in fourth-dimensional forms of vortexes or, as a comparison, the numerous cells, organs and other composite features of the physical body. Any one of a series of shocks can badly malform certain psychic structures. Therefore, the individual reflects outwardly

4

into his material life, the various mental and physical aberrations, many of which are puzzling our present-day savants who are trying to find cures for them.

All mental and physical conditions, without exception, are the direct result of malformed psychic structures. Call it what you wish, negative thinking or victim of circumstance, but you will find also that in the majority of cases, such aberrations had their first causation in some experience which was incurred in some remote and previous lifetime.

The problem or object here is to rectify the malformed energy structures in the psychic body. I have also found, just as did Jesus and many others who were in this work, that all conditions—mental and physical—are complicated by external forces which are called obsessions—devils, by Jesus. The subject of obsessions is tremendously vast and differs considerably with every case. There are actual entities from subastral worlds, obsessive thought forms, subjective circumstances, and many more which are energy creations. In all cases, a corrective therapy entails first going back clairvoyantly into the individual's past lives and finding the basic causes or circumstances of psychic shocks, which are the originating sources of trouble, and conscious participation by the mind.

The use of highly concentrated and directed thought energies, which is called psychokinesis, is all of the utmost importance in neutralizing or rectifying these negative energies. A subconscious, or sleep preconditioning, takes place and, above all, the valuable help and assistance of organizations of highly developed spiritual people who have advanced into the higher realms and dimensions, who are combined to aid in this work. This means God's wonderful and intelligent nature is aided and abetted so that not only are the conditions and obsessions removed, but the patient is reinstated upon a healthy path of evolution.

We are most interested in the healing and adjustments of such students and will appreciate your confirmation of the correction of any present existing conditions, or whatever your particular problems are. Rest assured also, that your help from us does not stop here, for our energies shall be forthcoming and directed unto you.

May you find new awareness and great joy in your adventure in Truth. Happy arrival!

Dear Student.

We are in receipt of your letter wherein you have listed a number of complaints which are indicative of psychic maladjustment, misalignment and general lack of knowledge on the facts of life. We are quite naturally compassionately understanding of your position and temporarily, at least, we will assist you through positive energy projections. In beginning your spiritual therapy, however, it seems to be indicated that our position and that of Unarius is not quite clearly understood. Our sole aim and purpose is to teach people how to lead and live better lives. It is unrealistic for anyone to assume that he can be completely healed of all physical and mental differences in a few short weeks of contact with Unarius.

It must be remembered that the little pit of clay in which you now find yourself took you thousands of years to dig. Conversely, it will take you thousands of years of evolution before you can dare to hope that you will be living without these psychic bonds which bind you to the material worlds, for the very fabric of your life is woven from these many past lives.

Spiritual healing is an age-old escape mechanism and engendered from unrelieved psychic pressures and is frequently exploited by various practitioners, cultists, and religious factions. In an exaggerated form, it assumes the diabolically conceived concept of divine intercession. Actually, if a majority of people living today were to be suddenly relieved of all their difficulties and problems, they would be completely lost, for in most cases, in common psychology, these various difficulties form not only a point of pivot but can also be considered safety valves or escape mechanisms.

Spiritual healing is all and more than the name implies. By learning of the evil ways of life through indirect reflection from the Superconsciousness, the individual naturally longs for a better way. This better way never has been, nor ever will be, achieved in the flesh in a material life. What happens in the way of spiritual healing to any individual living in a physical body is always the direct result of that person's learning how and having had this spiritual healing while living in the spiritual world prior to his present earth life. It must always be remembered that so far as the implications of karma are concerned, they are worked out only in succeeding earth lives as the individual progresses, learns, and is healed as a direct result of the preconditioning while in the spiritual worlds.

There are temporary measures which, when instigated in daily life and under specific conditions, can bring about what is termed spiritual healing; however, in all cases these temporary healing measures, while always the result of the spiritual preconditioning, are just that—temporary—and conditions, physical and mental, can constantly recur in many and varied forms until the person has developed the necessary wisdom and knowledge in the spiritual worlds and a closer relationship in his daily life with the Superconsciousness. When these conditions are met, and only then, can permanent correction take place. There are no miracles.

Belief in miracles is a misconstrued concept concocted by people who do not understand the facts of life. As it has been stated, it is only through the evolutionary process that we attain the necessary wisdom which makes these so-called miracles possible. Each individual is, in each succeeding life, pitting the knowledge he has gained in the spiritual worlds against the sins and iniquities which he has incurred in the physical.

Man changes but little in the mortal flesh, and only in such ways as he has determined for himself from the spiritual side of his life. Except in isolated instances, to comparatively advanced souls and to those who can attain the correct cycular junctions, these few seemingly hurdle the barriers of time and space and often accomplish more in a few years than the average person can attain in many thousands. But such people usually travel in the company of Avatars and Masters.

To the truth seeker, then, and one who is burdened with distressing inflictions of psychic shocks from past lives, be patient. Mastery over self (and these conditions are part of that self) is not attained in one lifetime, nor in many hundreds, for this is the basic plan which the Creator has for all of us—that in learning to conquer these material worlds, we learn mastery over self.

This, then, is the way of life: a challenge to each person that he will learn in time to develop from the selfish, personal self-ego and its selfish materialistic attitudes, a universally-minded being, attuned in all ways to the Creative Infinite Life Force, wherein this individual will oscillate into the infinite reaches and in all dimensions, and thus be a unit cell with the Infinite Mind.

Through Unarius, the student will be given necessary knowledge which will help him take the first and most significant step toward the better life, and most important, at this time he will be aided by healing energy projections to achieve certain junctions with the Higher Spiritual Centers of Unarius, wherein he may have previously achieved the necessary preconditioning elements which will directly result in cancellation of various negative blockages—true spiritual healing thus being effected and so starting a new evolutionary cycle into a higher state of consciousness with the Infinite.

(A letter from the Moderator in reply to a new student requesting healing of various ills.)

Dear Seeker of Truth:

Your letter was just received and inasmuch as you are no doubt in distress, we have therefore placed your name on the urgent list.

In reading through your letter, however, you are evidently carrying some false impressions or misinterpretations about Unarius. Therefore, in the interests of establishing an understanding of harmonic relationship, we would like to correct these false impressions and get you acquainted with the nature of our work and purposes. We are enclosing with this letter some other literature which, when you have read, will enable you to fully understand, so as to meet the necessary requirements which must be effected before you can be helped or healed.

As you will see, Unarius is not a religious organization—it is a true science, but much more than a material science—a science by which man will live thousands of years from now, just as people are presently doing in other higher planetary systems.

We do not pray, in fact, we are teaching and proving that all present-day religions are false. Jesus said, "Do not pray as does the heathen, in the street corners and in public places, or the synagogues, but rather, seek ye the Father Within (the closet of thine own mind), that He may reward ye openly." In our teachings, the Father Within will be fully explained; what this Father really is and how He functions; how to use the mind as a radio to keep in tune with this Father so that these wonderful constructive forces can flow into your body and into your daily life. You will be taught this and much more. When your body is asleep, you will be taken to such teaching centers as are on the planet Venus, and given additional knowledge and corrective therapies.

In our files are testimonials of thousands of people who have had wonderful miracles happen to them; all known conditions have been corrected. All this has been done without a single mind projection; in fact, you will begin to feel better even before you have finished reading this letter.

This is true simply because Unarius is comprised of untold thousands of more highly-developed spiritual personalities and of numerous professions (scientists, doctors, etcetera), all working from these higher dimensions of life which we call Unarius.

We are in full accord and with compassionate understanding in your need for healing and will say first that from the first contact with Unarius, the fullest and strongest possible healing energies were directed to you, and what relief and adjustments you may get, will primarily resolve into your conscious realization. In other words, by your faith be it added unto you. Ultimately, however, you will be expected to "pick up your bed and walk." As Jesus said, "Physician, heal thyself."

In these self-healing aspects, Unarius differs from any other religion or truth dispensations, for the prime purpose here is to teach the individual first how to heal himself, and with this wisdom will come the second additive which is a preventative and helps keep him out of future trouble, or should circumstance arise, to immediately correct the situation.

To properly understand healing, we must first become thoroughly familiar with the concept of the psychic self, your own true spiritual body which has been built up in that regenerative process through countless lifetimes. This psychic self is pure energy—energy as it is found in any dimension—and wave forms which are thus shaped by the particular experience with which they are associated. These countless billions of wave forms in the psychic self reside in a concentrated amalgamation of vortexes somewhat

11

analogous to the various organs of the physical body. Even more than this, however, they are actually responsible for the physical body and every process of regeneration which is concerned with the building of new cells, et cetera. Idea and form have been carried over from one life to the next in these tiny wave forms. The same procedure is carried on in the mental processes; that is, the conscious mind is only the objective point of focus from a certain portion of the psychic self which can be called the psychic mind, or as psychiatry terms it, the subconscious. Here again, in the experience of molded wave forms, an infallible portrayal of past experiences is being constantly reflected into the conscious mind in an oscillatory condition back and forth.

The psychic self, however, has one more dimension which is the superconsciousness or Christ-self. In the physical brain of a human, the back portion (cerebellum) of the brain controls all the involuntary muscular reactions of the body. The frontal lobes (cerebrum) are responsible for the voluntary motions of the body. The two portions are somewhat loosely linked by a gland-like portion at the center base of the brain called the hypothalamus. In the psychic mind, the same separations exist; the subconscious is now the cerebrum, the super or Christ-self, the cerebellum, and the hypothalamus is actually the linkage to the Infinite.

All of these things, of course, must again be conceived as wave forms and vortexes and their subsequent radiations, which are actually an aura which is more properly called harmonic wave forms or vibrations which are compatible in frequency to the dimension in which they function. It can now be easily understood that our present life is an exact portrayal of previous emotional experiences, the only difference being in the time and place. If we are to dissolve or rectify these unsuitable negative conditions, we must remove the negative forms from the psychic self and replace them with such wave forms which reflect a more suitable and acceptable dimension of interpretation in our present life. This is cause and effect. It does one little good to have these conditions even temporarily removed by some great healer or healing which acts only on the exterior physical self. Unless the wave forms are changed in the psychic self, the individual will soon find himself back in the same negative dispensations.

Contrary to general belief, no Master or Avatar can change these wave forms for you, for only you yourself can do so; for should any great Master be able to change these wave forms for you, he would be violating the cardinal principle of life itself, which is that every individual, if he is to become intelligent, must first learn of Infinity, and to do this he follows through in the evolutionary pattern, an infinite number of experiences or emotional lifetimes until he has learned of this Infinity and how to conquer or properly utilize experience in a constructive way.

If a young man is to become a director or high official, he would be wise to start at the bottom of the ladder in the factory and learn about the entire workings of the organization from the ground up. This is exactly what God had in mind in the creation of man (if we can personalize God for clarity's sake). Another way would be to say God is infinite, is the sum, substance and total of all things and is infinite only because He is finite in all things and that man is the ultimate resolution, spiritually speaking, of the infinite form and mind of God. God assumes in every individual finite personality and again Infinity.

Therefore, no Master would change the tiny wave forms of experience in your psychic self for he knows that these are all that you are, and that to change them would be tantamount to spiritual murder. Instead, the natural and only way is to change these things for yourself. Inasmuch as it took you thousands of years and many lifetimes, and realizing this, do not expect them to be changed instantly, for even if it were done, you would be like a fish out of water; you would find yourself totally unsupported in any idea, thought or form in your daily world. Instead, expect these changes called healing, to take place in a more normal manner, constantly rebuilding new cells and tissues. You must begin to replace the old negative wave forms in the psychic self with more constructive, positive wave forms which portray the more desired life in the future.

Look at this future as a challenge to your strength and knowledge, not in terms of days or years in this lifetime, but a future which stretches endlessly beyond the eons of time or even eternity. Only in such concepts of positive interpretation of life will you be enabled to live in that future. When once you have started to become master of your destiny, you will no longer be mastered by the past. When this principle is thoroughly imbued and consciously expressed into every moment of life, you will begin to be healed; you will see day by day your life changing and just as rapidly as you wish it to be changed. The old negative forms will slink away like misty shadows before the morning sun. Your life will begin to take on the shining luster of the Infinite Intelligence. Moreover, this will be a magic alchemy which will radiate into the lives of those about you causing in them the desire and awakening.

Thus you will free yourself from the chains of materialism. Like the fledgling of the lark, you too, on new-found spiritual wings, feathered with Infinite Wisdom, will be enabled to soar off into the limitless vistas of eternal life to become unified with all creation. Nor shall you return to the old pitfalls of the material world, but always remain a shining lustrous beacon of illumined light, to reach out with your rays of love into the stormy seas, over the distant horizon into these material worlds to light the pathway of the storm-tossed mariner on the seas of life.

When you begin to understand all this and much more—all of which is explained in our various texts—your life will change, your condition will become just a bad memory. However, we must be fully understood on this point, as we are teaching a permanent, self-corrective knowledge of life. The results you attain will depend largely upon your understanding, your dedication, and your purpose to become a better individual so that you may unselfishly be a better servant to your fellowman.

We therefore welcome you into Unarius and irrespective of what action you may or may not take in replying to this letter, we will nevertheless do all we can for you.

We are always gladdened by an opportunity to share our better way with any and all who ask, regardless of whatever conclusions our contact may reach. We know that we have all been benefited and made better for having so joined in the common union of faith.

Projections of the Radiant Energies unto you.

A Letter to an Ill Student on Healing

Dear Student: We have just received your letter giving us up-to-date information on your mother's condition. Inasmuch as this letter is, in essence, typical of some which we have received here at the Center, may I objectively pass upon some concepts and principles involved in corrective therapies or healing which are universal and applicable to any and all humans. They are highly scientific in nature; in fact, the actual science and its works are generally called Infinite Intelligence.

As students of Unarius, you have some prefaces of these concepts and principles; however, your letter clearly indicates that they are not properly or thoroughly understood by yourself and your mother. With this and associated factors involved, healing cannot take place until such time when they are more properly understood and usage begins to occur.

There is strong indication in your letter that you and mother are still clinging to old Babylonian and pagan religions of which you believed in the past. All religions, including Christianity and their associated forms of expression, such as prayer, are pagan. There is a much better way to contact Infinite Intelligence than through prayer. This is through constant conscious attunement, which is automatically brought about when the Infinite Intelligence has begun to be understood. Through this conscious attunement, the Creative Intelligence can and does institute a much greater degree of corrective therapy and in proportion to the strength of your attunement and understanding.

Contrary to what you may believe or may have read, there are no miracles, nor have any ever been performed. This may sound extremely radical to you, but it is entirely provable, and it is hoped that you too will understand in the future, just as we do. Miracles in the physical world and the physical body are brought about only after certain spiritual preconditioning takes place and which has, in effect, already healed the person psychically before the person was healed in the physical body. This is most necessary to give the person involved the correct balances in retrospection between physical and spiritual worlds, thus aiding his evolution.

Men like Jesus were only instrumental in bringing about what was seemingly a miracle because they could objectify to the person being healed in the physical plane, what had already occurred in the spiritual. This objectification took place in the form of a certain catalytic power or energy which was projected psychokinetically and which momentarily transcended the sick person, breaking the chain-like oscillating reaction between the diseased body and the existing, remaining energy malformation in the psychic anatomy.

When this reaction was broken, the psychic malformation was rectified or changed. The diseased condition in the physical body, not having its parent oscillating point, disappeared. Putting it another way, look upon every human being not as a physical body but as a beautiful rainbow-hued creation—which is the psychic anatomy. It is pure energy, made up of billions of little waves of energy, all going around in their respective vortexes and repeating endlessly the same experience which originally caused their form and shape. Conversely, they are affecting the outer shell, which is the physical and which is actually only more energy in the form of tiny atoms.

To affect the physical, we must first produce the effect in the psychic anatomy. That can only be done when the psychic anatomy is in complete accord with the objective to be attained. That is done through various wave forms of energy which come to it from numerous sources. Some of these come from the physical body and which are always connected in an auto-suggestive fashion to all of the past experiences. They will consequently obtain rapport with that part of the psychic anatomy which was fashioned, so to speak, from the past and produced a reaction known to most people as thinking. However, the psychic anatomy receives energy waves from many spiritual or astral worlds in just the same manner as a television set receives pictures from the transmitter. Some of these Spiritual Worlds are very beautiful and create a strong desire for this psychic-anatomy person to ascend unto them—which it cannot do because it cannot oscillate or participate in them by reason of the difference in frequency rate.

These Higher Worlds must therefore remain in the inner consciousness or psychic anatomy of this person as a vision which gives the person a sort of libido or drive, or attainment. When this condition comes about, the old past becomes odious in comparison, just as one might look at a beautiful new hilltop home in comparison to a dirty hovel which he now occupies— even though in his childhood that same dirty hovel was his whole world. It is, in such a comparative position, that the past and its many negative shocks and indispositions are, in effect, intensified and oscillate or are reflected into the physical body as different kinds of disease.

By now you should begin to see that miracles have, in their physical sense, never actually happened; neither can all the praying which you or anyone could do could change them—no more than the doctors can. Only in a transcended moment can such a happening as a so-called miracle take place. The visions of higher worlds, the various associated preconditioning and cancelling actions which have taken place in the psychic anatomy will, in this transcended moment, enable the complete cycle of healing to take place through the physical anatomy.

The principle and processes I have described to you are inviolate. There are no exceptions. All past stories of sorcery, witchcraft, divine healing, such as was depicted in the New Testament, are only stories out of the past which have been derived from thoroughly misunderstood, misconstrued and distorted fabrications which may have happened historically or otherwise.

The Infinite Creator functions in principle—immutably and inviolately. No person can aid in the so-called miraculous healings until he understands and becomes part of this Infinite Intelligence. As part of this Infinite Intelligence, he functions just as inviolately as the great Creative Intelligence—a function which is predicated upon a progressive evolution and which, in the individual human, can conceivably make this human into a master or godlike person if the evolution is carried forward long enough and far enough.

And so, dear student, do not look for someone or some god to perform a miracle for you or your mother; instead, look forward to the day and the hour when, in understanding the great Creative Intelligence, the miracle will have already happened. You—and you alone —in the knowledge which you have obtained, will have worked that miracle for yourself, and by that time, you will be far beyond the vale of earthly tears, never to return.

For the immediate future, you may be assured that you and your mother are under the protectorate of the great Universal Unarius where millions of dedicated souls will be constantly radiating the power of their Intelligence to you both. How well you will receive and be benefited depends entirely upon you, individually, and how well you can attune yourself unto the Higher Worlds in which they live.

Try to detach yourself, from that overpowering sense

of physical importance. Look upon the body the way you would look upon a boil or a festering sore, because that's all it really is in comparison to the higher way— a festering sore which is the sum and total of your past. You will live in a Higher World only when you have learned about them and have the knowledge to live in them. By contrast, this now very important physical life will have become a dirty, filthy, pain-racked pigsty remembrance. As a person who knows about these Higher Worlds and can describe them, scientifically prove their existence and prove in many other ways, that I have come from one of these Higher Worlds for the express purpose of dispensing this knowledge to those who carry the inner vision and so make the vision a reality. I can truthfully say that all I have told you is true and there is much more which I cannot tell until you have the ability to comprehend. Yet it must remain with you to prove the truth and the efficacy of this truth in changing your life.

I am looking forward in strong faith and anticipation that you and yours will be among those happy pilgrims making a progressive evolution into eternity.

(The following was voiced by the Moderator in reply to a student with questions pertaining to Unarius being similar in principle to other present-day schools of thought.)

Dear Friend, from time to time, (fortunately, not too often) we here at the Center receive a letter from a person who is inquiring about Unarius or perhaps sending for the first book or lessons, and in this letter the new contactee makes references to certain spiritualistic or cultistic groups or organizations wherein it will be pointed out, the contactee has certain teachers or "guides", or that there are other relevant connections and experiences in these spiritualistic affiliations. Your letter is a typical example.

Before organizing Unarius, I personally spent some fifteen years among the spiritualists of California, and which included from time to time, contacts with groups or mediums throughout the entire world. I was not just a mere observer, I was an ordained minister in several of these churches and still carry papers. I gave lectures, message services, et cetera. I did everything these mediums could do and a number of things they could not. I gave lectures on subjects about which they knew little or nothing. I gave messages which they could not duplicate and, on one occasion, I stopped one medium right in the middle of a sentence by a silent thought projection and she could not speak until I released her. I have demonstrated teleportation and telekinesis by appearing in two or more places simultaneously, by starting inoperative television and radio sets to working sometimes miles distant, by causing different kinds of inoperative mechanical and electrical equipment to function properly, such as generators, voltage regulators, etc., and many others.

I know the people you write about in your letter; I've

been in their church many times, and I believe I'm the only one who ever challenged the individual in his own church and proved his teacher or guide a phony before the congregation. Also, I gave messages and readings on street corners, in buses, cafes, dance halls, as well as churches, which no medium could do; sometimes done in broad daylight. I made positive and provable contacts with deceased persons in other worlds, some who had been dead for many, many years, even in bright sunlight. I located lost objects, uranium fields and oil wells, thousands of miles distant. I drove cars, guided people's hands and other types of phenomena, as far as two thousand miles distant from the scene.

It should also be noted, whenever contacts were made—readings, telephone, mail, or even indirectly—such contacts always had immediate effects. Countless thousands of persons have been healed from innumerable kinds of conditions; cancer, fistula, dropsy, crossed eyes, impetigo, diabetes, virus, flu, hay fever, asthma, headache, smoking, et cetera. Countless others have been temporarily transcended, felt overwhelming heat, and not a few have been antagonized by an unknown (to them) power. All of these feats can be proven. I did all that without guides or teachers, without trance, or without any of the mumbo-jumbo commonly practiced.

In our files we have many thousands of handwritten testimonials of miraculous healings of all kinds, most of which are much more miraculous than those described in the New Testament. We have thousands of eyewitness accounts of spiritual phenomena that cannot, nor have ever been duplicated by any spiritualists. When you know Unarius better, you will read much about this, and perhaps if you are dedicated, and you can put aside the false convictions and spirit practices, you will have your own phenomena; your life will be changed.

Unarius is an organization of Advanced Beings from

a world which is far above and beyond any astral planes contacted and worked with by the spiritualists. There are millions of these Advanced Beings, (of which I am one). They do not have names and do not wish to be called by such. They work silently, without thought of praise or personal flattery. Their sole purposes are for people like you, to try to help you climb out of your earth-world of pseudo spiritualisms, religious superstitions and other witchcrafts.

The Unariun Brotherhood is a scientific organization and teaches true Creation, together with all known human aspects which relate man in the total Infinite Creation. Unlike any spiritualistic group, our books are in libraries and teaching curriculums of some of the country's largest universities. And in these books we teach what spiritualism really is. Even though I have and still can, out-demonstrate any spiritualist on Earth, without exception, I would not like to be called a spiritualist.

To me, a spiritualist is an ignorant person, doing something he or she knows nothing about. He is totally unaware of the science involved in spiritualism, and in this pseudoism he has made himself subjective to certain hypnotic malpractices, and he will likewise intimidate and ensnare all those who listen to him, in the same false hypnotic malpractices, all of which should be placed under the common heading: "Escape Mechanisms". An escape mechanism is always indulged in by a person who cannot obtain an objective, intelligent analysis and subverts his life by an escape mechanism. I could go on for hours on this subject and I can prove all I say, which no spiritualist can do. None of them can prove the validity of their supposed teachers or guides or their supposed contacts!

So once again to you, or to anyone in a similar position: your one hope, your one chance of escaping the hellish nightmare of this rondelet of false pseudoisms is to become intelligent. Start to learn Creative Science, and in Creative Science, you will find your own place and the answers to the enigma of your life.

Now, if any or all of the above statements in reference to my accomplishments seem immodest, I will repeat an old cliché; "The ends justify the means." Anyone who knows me could testify that I am a very modest person in all ways, manners and means. Besides, the above statements when compared to the total of all accomplishments would be extremely modest.

It should also be considered that since entering into the active propagation of Unarius, we have been assaulted by a constant and never-ending flow of letters from inquisitive hopefuls, aspirants, new students and others who, without knowledge of our intents or purposes, have, in their own judgments, classified us as some new cult or theorism, readapted religion, or even one of the more frenetic howlings in the wilderness of retrograde occultisms or theosophies, or so-called mind sciences.

Such effrontery could not reasonably be expected to go on indefinitely without challenge. Ruth and I have therefore begun to take stronger measures to keep Unarius properly classified as a true Science, an expotnential science which constantly expands the intellect, squared in interdimensional ratios, to a point well past any temporary equilibriums, seemingly established in the mind of any person who, in reality, is feeding himself on the mildewed fodder of an antiquated religious superstition, spiritualistic cultism and other placebos, invented for the purpose of a contrived escape-mechanism figurehead.

Immodesty is only a comparison established by certain parallels. So far as the Mission with the Unariun Brotherhood is concerned, there are no established parallels and no comparisons. Unarius is unique; nothing which could be compared to it or similar, in

24

any aspect, has ever appeared on this earth—save the Mission of Jesus, the totality of Amenhotep (Akhenaton), the sciences of Anaxagoras, all of which, however, are not comparisons but adjunctive efforts.

As of today, Unarius achieves the totality of its Mission in the atmosphere of our twentieth century technocracy; it has achieved an expressive redundancy not possible in any other age or epoch. Yet, Unarius is much more than a classical or interdimensional science. It is a demonstrative science, aided and abetted by millions of Advanced Beings from a Higher World.

So clear away the confusions and cobwebs from your mind; clear the waters of introspection and let the beauty of Heaven be reflected, at least long enough to make objective comparisons. The hide of the rhinoceros or an elephant is supposedly thick and impenetrable, yet is but a gossamer tissue when compared to the hide-bound opinions which have been fostered and nurtured in the darkness of an ignorant mind.

(The student to whom this letter was a reply, asked several questions regarding the Science.)

Dear Unariun friend: Now about those questions. In themselves, questions like these indicate a person who has or is trying to go beyond the commonly associated patterns and levels of earth society—a person who is attempting to begin the process of evolution which will eventually, in the eons of time ahead, metamorphose him into a spiritual being.

It's questions like these that have been asked thousands of times by thousands of aspirants and that's exactly what Unarius is all about—to answer these and countless other questions which constantly arise.

In its total curriculum, the Unariun Library is complete and will supply answers to any and all questions for many thousands of years. We would only be doing you a disservice by trying to piecemeal, as it were, bits of information which might seem relative at this time, to your questions. The proposition is, by beginning to read and study, by constantly pursuing this never-ending quest, you will, in time, begin to understand.

There are many levels of consciousness to be attained through reading the books and lessons. They are a never-ending source of information. New clarity, new answers appear as a new level of consciousness is attained. And don't forget that wonderful transcending, life-changing power which is always projected whenever a student picks up a book and starts to read.

The concept of Unarius as a Mission is relatively simple. Countless ages ago, millions of Advanced Spiritual Beings, living in the timeless-spaceless Inner Dimensions, were able to look into future history of the earth world and to very accurately plan what might best be described as a life-saving Mission. They realized that without certain aid and assistance, the peoples of such an earth world could only advance in their evolution to a certain point—and as all progress must be constructively biased, these people must be helped in the jumping-off process which would eventually lead them into one of the inner dimensions or kingdoms where they would be not only self-supporting, but would be sufficiently constructive to aid in the grand plan of creative evolution.

And so this Brotherhood would, from time to time and at an appropriate period, send one of their numbers as an Emissary, so to speak, where, more directly attached to an earthman-body, the Emissary would be able to propagate into the earth world the proper part of that particular mission. In a relative way, this is analogous to an ordinary television set which is tuned to a certain station except, of course, the program material is vastly different and also that it radiates out into the earth world many thousands of different kinds of energizing, healing, and transforming power-beams.

We might enumerate a few of these Emissaries: Zoroaster, Amenhotep, Anaxagoras, Jesus, et cetera, and in certain ways, they were all one and the same; they all relayed into the world at that particular time, a certain preparatory phase of the final mission which is now being propagated.

It is imperative, for your own sake and for whatever hopes and aspirations you may have in the future, that you learn about the grand plan of creative evolution. To survive, you must eventually develop yourself into a spiritual being who lives without a physical body in one of the inner dimensions that is not only entirely different from this earth world but which would defy description and the necessary intellect to comprehend it at this time.

Therefore, be patient. The questions you now ask will be answered, and you will ask these same questions in different ways in the eons of time in your future, as you seek new answers, for this is indeed your future—a future which is not wrapped up with a web of conflicting ideas and in a neat little box, as it were, wherein most people hope to justify the entire compass of their lives. Yet, as they live from within this little box, so do they ever spin, from day to day, new and tenuous strands of conflicts and eventually, a cocoon so strong, they would perish but for the way in which the power and intellect from a Higher World would help them unravel this tissue of their past.

Please do not misunderstand us and our position. We are not a Messianic Mission as it is associated in Christian beliefs. We do not pretend, nor do we promise to save anyone because they believe in us—or anyone else, for that matter. It is a mandatory requirement that you must change yourself. We supply the proper tools, (actually, knowledge) along with the necessary inspiration, demonstrations, and judicially applied corrective healing powers. It is quite senseless to believe anyone can be saved—which means, in effect, to be suddenly levitated into one of the inner timeless-spaceless dimensions, which would be absolutely incomprehensible without the necessary evolutionary preparation and growth.

In its final phase, as it is now being propagated, Unarius is the Second Coming; the Emissary or Moderator is, in certain aspects, a compound of those certain Emissaries who have gone before and, in particular, Jesus of Nazareth. Indeed, the mission was so appropriately timed, with the coming of the atomic age and the explosion of the thermonuclear bombs, wherein the Biblical prophecies, such as were given by Peter, were fully and adequately justified—that He did return at the time when the heavens rained fire, and even the elements burned with the fervent heat—and at that time, to begin a lifesaving Mission among the hundreds of millions of human derelicts, scorched and burned, as it were, by the vast dross of human materialism—a swirling maelstrom of endless and repetitious conflicts wherein all seemed to be sucked down into some vast, endless, darkened depth. These hopeless ones' only hope was in the way in which the Unariun Brotherhood could, from the inner dimensions, enter into the secret closet of their minds and there light the flame of life—a Light that would later brighten the darkened path which would lead them into the realms of light above.

So let it be, if you are one of these who have, in your secret closet found your light, then let it lead you to where we await.

Dear Student: Occasionally some student letter makes reference and implies certain comparisons to one of the currently existing forms of religion or metaphysics. In making a direct reply, we will therefore relate a letter which was sent to a Unariun who made a similar inquiry:

It is most important to stress, that as of this time, there are no religions or metaphysics which can be compared to Unarius; even if any of these religious or (so-called) metaphysical or mind sciences have included in their various liturgies, certain scientifically sounding references such as vibrations, cosmic rays, atoms, or a host of such similar nouns which imply a scientific meaning.

These various terms have been merely borrowed to "dress up" a very unscientific religion or mind science, with the intent to make it more impressive and give it that certain scientific authority. The use of a few scientific words or terms does not make a religion scientific; moreover, the persons who use these terms or expressions are not trained scientists and do not understand them; also such terminology is always unrelated, has no continuity, and proves not a thing. It only impresses the gullible ones who likewise do not understand.

Obviously, as the world is at this time quite scientific, various religions and their re-adaptations are, by contrast, even more pagan, barbaric, and unfactual, and to survive into the future they must be either completely reformed and readapted or they will surely pass into the museums where they most properly belong.

Do not be misled; if man can, in his third-dimensional science, strike certain parallels wherein he has created the many marvels of this decade, then surely he has unwittingly instituted the same creative processes whereby the Infinite Intelligence created all things.

To be religious means a factual intelligent relationship to the Infinite Intelligence whereby a complete, constructive evaluation of Creation can be clearly visualized, and can be seen to be the existing and re-creating form and substance of the material world, and also, the same creative evolution is sustained in higher, commonly called Spiritual Worlds.

Two thousand years ago Jesus explained this as the Kingdom Within. The so-called religions of our time have made a barbaric farce of this great man's teachings. It is the intent and purpose of Unarius to reestablish the teachings of Jesus and in the form and manner of our present third-dimensional science, where it will be sustained by irrefutable evidence.

Therefore, if you have made comparisons with Unarius and other religions or adaptations, and you have found similarities, then you have not as yet gained a comprehensive evaluation of Unarius. In the various books and lessons, there has been frequent reference to these existing religions and adaptations. However, this was always done to point out the obvious disparities which existed between these various religions and a more true scientific understanding of creation, and with the exception of the more esoterical values of the teachings of Jesus, no inclusions have been made as part of the correct fundamental teachings of Unarius.

Even though our present-day, third-dimensional science is fundamental, it, however, does not go into the fourth or adjacent dimensions, and while science recognizes these other dimensions, it has yet to include in its precepts, the exact knowledge of what these dimensions are. This is the most important part which Unarius plays at this present time; to start the beginning of a new interdimensional science which will develop in the hundreds of years in the future.

You, dear student, have a great advantage; through Unarius, you can take a quick shortcut into the future and gain the knowledge which this future mankind will have—and most necessarily so, to sustain his earth life. Either that, or he will reverse his cycle and through the potentially destructive nuclear power, plunge the world back into some primitive beginning.

Celestial mechanics is an exact science and is absolutely impersonal because it is the Infinite Intelligence and Creator, which is the sum-total of all things, as well as the Creative Principle which created these things. To understand the Infinite Creator, we must first abandon the old earth world ideosophies which are nonfactual and nonscientific, and begin to relearn life as the Creative Principle, not as a dominant deistic configuration, but as an exact science which is completely and absolutely impersonal.

I might warn you, however, that there are countless millions of astral underworld entities who are using the weaknesses of people in their various beliefs and idolatries to impede a more natural evolution or even to parasitically control them by erecting some false image in their minds such as a Christ or Supergod, et cetera.

It is therefore best for your own safety and welfare to completely abandon and destroy these age-old ideosophies from your mind and begin to learn interdimensional science as it is postulated in the books and lesson courses, and to continue on this straight and narrow path, not being tempted by comparisons or tainted by contact with them.

As Jesus said, "Do not put new wine in old bottles, lest they burst." Therefore, in our study of Unarius, let us completely dissociate this complete interdimensional science with any of the age-old systems of idolatries; they are unscientific, unrealistic, and will lead you only deeper into the mire of karmic despair.

Every human must learn of the true Infinite Creator and the Creative Principle by which the Infinite Creator lives, and in learning of this Creator and the Principle, this human will then again express a supreme effigy of Creation, thereby becoming a part of this Infinite Creation, but not as some false god, supposedly ruling mankind through force, coercion, fear and superstition, and, most certainly not by exhibiting a malevolent attitude; neither by common human frailties, nor the more destructive potentials such as murder, revenge, violence, lust, hate, fornication and various coercions which are commonly portrayed in various religions in their deistic interpolations. Such concepts are found only in the minds of people living in the astral underworlds and most certainly no human should pin his hopes of a better future by being subservient through fear to such a god configuration.

To attain immortal life in a Higher Spiritual World means simply this; you must become a part of it, just as you are a part of your present day world, and no one can live in any higher spiritual plane if he is ignorant of the way of life which is lived there. Life in a Higher Spiritual World demands that you know of the principle of Infinite Creation and how to use it in your daily life.

There are no short cuts to this higher way, there are no substitutes, there are no saviors nor messiahs who will place you in one of these Higher Worlds, for they would surely know that it would mean your death to do so; for you would surely die in an environment which you did not understand nor could you comprehend.

Your evolution into the future is the sum-total of your life. Do not let anyone rob you of this future by presenting a false promise, even if it is all dressed up in some glamorous or supernatural aura. The robes and vestments of some priesthood can be your shroud, and the neighborhood church can well become your coffin should you choose to place your future in the hands of those who have constructed these artifacts.

Therefore, in the future, as you begin to learn of the true Infinite Creator and the Principle, learn to keep positively attuned to this constructive Principle. In this way then, you will begin to let the Father which is within, start doing all things, and which can quite naturally be considered to be constructive. In keeping positively attuned to the Creator through the Principle, you will be finding the Kingdom which is also within; you will also thus keep yourself correctly placed and morally responsible for all thought and action and your outside, material world or earth life will begin to be a manifestation of the Kingdom and the Father who lives within.

This is the only way in which you can begin to start your flight into the Higher Spiritual Worlds.

Dear Friend: Occasionally a person or a new student who has just contacted Unarius and after having read "The Voice of Venus" may write to us: how is it possible Sha-Tok or Jesus could be living in a Higher World and still be reincarnated on the earth at the present time? This is, of course, a misunderstanding and has been quite adequately explained in different places—bonus sheets, et cetera, in the liturgies.

It must be thoroughly understood that a highly-developed Spiritual Being, like Sha-Tok, together with his oscillating polarity Erza, as I have described it, could not possibly live on the earth. It has been many thousands of years since such a Being could live in an earth body. The difference in the frequencies of these energies, the tremendous power, et cetera, would literally consume an earth body in a matter of seconds. This Being does, in a sense, however, live in an earth body, but in a manner similar to a closed circuit television. Just as it was in the past with Jesus, Akhenaton, Anaxagoras, Spinoza, Swedenborg and a number of others, a certain expressive continuity was maintained in their respective idioms in order to best serve the planned purposes of that particular Mission.

Now, when any of these personalities died, certain portions of their psychic anatomies relative to their earth life were stored much like blood is stored in a blood-bank, except that in the case of the psychic anatomies, this is energy oscillating in closed cycles fourth-dimensionally. The Advanced Beings in the Higher Worlds can do this storage work much in the same manner as putting food into a refrigerator.

Sometime before I was born into this earth world —that is, speaking of the physical body—a psychic anatomy was constructed from these "stored parts" which formerly belonged to these personalities I have above mentioned. All of these parts were directly concerned with making and regenerating the physical body as well as the different mental processes, etc. After the psychic anatomy was constructed, it was, at the proper moment, connected to the fetus at the moment of conception and began to reoscillate or project the necessary information of the cell structure of the ovum and sperm which, in turn, coded the DNA molecules, thus starting the building of the physical body.

Mental processes were likewise instigated, plus a direct oscillating beam or connection with the Being Sha-Tok, so that at suitable times throughout my physical life this Being, plus any other Beings, could project pictures and other kinds of information, plus any and all suitable kinds of power necessary to correct, heal, or prepare any conditions which I might encounter in my application of the Unariun Mission.

It can now be seen from this explanation, that my life from birth to my future passing is quite different than any person on the earth who maintains a direct continuity with the same psychic anatomy from life to life as he reincarnates. To a certain degree many people are affected or even directed by a number of astral agencies; in fact, all people are traveling, so to speak, with the combined influences of many astral worlds.

Only with the Unariun Organization, however, the highest and most beneficent form of health, education and welfare is being administered to those who, to some degree and in some way, have had certain preconditioning. Unarius marks a distinct line of cleavage with the third-dimensional earth-world place where all people live according to the more primitive laws of their beginnings.

As you will read in the Unariun liturgies, life in the Higher Worlds is very distinctly different and to get there is a matter of personal struggle to overcome the past earth-world lives, rebuild a completely new psychic anatomy or energy body which is done through acquiring knowledge in the educational process. I feel sure that you or any other person, devotee or student will become much more comprehensive in this re-education, rebuilding process through the liturgies of Unarius. Furthermore, these studies are always accompanied by that wonderful power which heals and adjusts as education, reorientation and cancellation of the past progresses.

Dear Friend:

In answer to the so-called "Ashtar" (or other such) messages, this letter is being written in answer to the many quests pertaining to topics contained herein. The situation mentioned before with the husband could apply to many persons on the path of Truth, as many individuals are endeavoring to "change the other fellow." As one of the first issues of your letter, we know explanations will be important to you, as well as many others.

You remember many years ago the story of Tarzan who, as a baby, was stolen by the apes in the jungle. His parents killed, he was then adopted by a foster mother, a great anthropoid ape. Growing up in his childhood years, Tarzan was unaware of the great world around him. The jungle with its animal and bird life was his entire world in which he played happily from day to day. At about the age of eight, he stumbled into the father's hut, which had been left deserted through the years. Through books he had found within, he learned to read and write. He learned much of the great world about him which he had not seen, so that when he became a young man, and a ship finally landed on his shore, he was quite able to slip easily into the civilized world. Picture, however, if he had been plucked from the jungle without this preconditioning and flopped into the civilized world, he might have become a hopeless imbecile through shock and fright.

This picture presents a portrayal of the position many people occupy in regard to flying saucers and other interests of interplanetary travel, as well as other issues which have become of widespread interest throughout the world at this time. If you could become sufficiently clairvoyant for just a brief time so that you could see all of the infinite universe around you, at that

moment, you would be like a Tarzan without any preconditioning, who was suddenly flopped down in a
great civilized world, and the shock of seeing this
infinite universe in this way would be too much for
you. You would see a terrible and awe-inspiring pandemonium of spacecraft, flying saucers, huge brightly-
colored planets, great and small vortexes of rainbow-
hued energies; this and much more, all seemingly gy-
rating, weaving, spinning, passing through, in and out
of each other, all apparently with the utmost confusion of a great maelstrom of strange and awe-inspiring
sights; unseen people living in and around each other,
around and in the world, planets and spacecraft passing through the world without a whisper or a breath.

Yes, this would be a sight which no mortal mind
could endure; yet in some future time and in some
spiritual dimension, you may be able to view all this
in just about that way. Yet then, you will be viewing it
complacently. You will know that it is not chaos and
confusion, but a well-ordered and well-regulated trans-
position or orderly procession of all the infinite things
and concepts within the mind of the Supreme Intelligence. So do not divert your way or thought from the
true course. A flying saucer addict or fan is somewhat
like the hillbilly who stuck his head out of the bushes
and saw his first automobile.

Another issue in your letter which goes along with
spacecraft is this Ashtar situation. Do not misunderstand me; I have nothing personal against E.P.H., but
I say it is indeed unfortunate that such a fine person
should become duped into becoming a tool for the
highly organized black forces. Ashtar is a clever device
used by these black forces which is intended to trap
and enslave countless thousands of people, turning
them away from their true course of evolution and
leading them into the quicksands of oblivion. Throughout the New Testament in the work of Jesus and his

apostles, there are numerous warnings about these latter days of false prophets and teachers. Jesus said, "By their fruits ye shall know them." Let us analyze this Ashtar. First, it is based on an old psychological concept known as an escape mechanism. Through the many and diversified pressures generated in this highly complex and civilized world, untold millions of people have become neurotics and are seeking a means to relieve these inner pressures and conflicts within their minds. This safety valve is called an escape mechanism. These various escape devices do many strange things to otherwise fairly normal people. You see people constantly portraying in thousands of different ways, various kinds of escape mechanisms.

The black forces have hugely capitalized on these highly civilized and neurotic times. If you think for a moment that it would serve or benefit anyone to be snatched from fancied oblivion and carried off into space in a ship—yes, ten times ten million space ships would not be sufficient to carry all the people away from this world at one time. Think, too, of all the panic, the pandemonium and confusion—like seeing a tremendous enlarged version of the sinking of the Titanic.

Millions of people do not know about, or believe in spacecraft; think of the shock and distress to them. How utterly unrealistic! Then after they were saved from this so-called cataclysm, what to do with them then? If the earth is destroyed, it is assumed they would be taken to another planet. There they would have to start life all over again, but would still be the same frustrated people, only now much worse off than before. Even if they had survived the terrible shock of moving thusly and had enough senses left to start life over again on this strange planet, how many years or generations would it be before they would again be in eminent danger of destroying themselves without completely changing the entire pattern of evolution which they had built up in their psychic structures, through many thousands of years. How would they work out their karma under such strange conditions?

No, indeed, Ashtar is another dream, a device used to exploit and condemn any poor unfortunate soul who may rest within its shadow. We never overcome by trying to run away or escape. (How much better it would be to recognize these escape mechanisms at work. There are various gimmicks contrived to divert man from his true evolutionary pathway. This so-called Ashtar astral entity, supposedly head of 10,000 spacecraft standing ready to "lift off" from the earth 144,000 chosen ones, is one of the negative factions to feed man's fears. You shall find it will amount to naught other than to enslave the minds of those who heed his messages.)

(This student had been partially and at times, totally obsessed and possessed by very strong negative and evil forces which sometimes completely controlled his mind and thinking, often to the point of having him believe he was some great master and that he was fully aware of all energy principles—ready to change the world! George was material for the psychiatric ward, often demanding exactly what the Moderator should do to help him, even though he, himself, claimed to be a master! It was shortly after this information was sent him that he experienced complete release of all his evil obsessions, leaving him a demure and humble man with gratitude overflowing.)

Since the beginning of our association I have taken more than the usual interest in your case because, for one thing, I know of the tremendous struggle you were making against various obsessive influences. Your letter of this morning, however, has inspired me to answer you directly (personally) because it is now clearly indicated that you have entered into an even more dangerous phase of your self-liberation. Your letter states that you now have a fair understanding of energy. My dear brother, may I say this is probably the overstatement of the year and one which many students make after reading the books and lessons. It must be borne in mind that these texts are tailored to fit the minds of those who are just beginning and have, up until this point, been completely ignorant of the most rudimentary knowledge of energy and the Creative Intelligence, which is composed of energy.

Therefore, these texts are not presupposed to be the ultimate conclusion in learning but only the first step taken by any student in breaking away from the false doctrines of the material world—a step which must be taken by all those wishing to continue into eternal life. Furthermore, Unarius is not just another escape mechanism whereby a person can escape the insecurities of the material world, for in presenting this constructive evolutionary knowledge, it is clearly indicated that each future dimension of life will have its own particular set of expressionary forms and these attendant demands that they be solved in constructive knowledge.

In this lifetime, I have spent some thirty-five years actively engaged in electronics, various systems of communication, etc., and it is safe to say that I spent many thousands of lifetimes in the direct pursuance of learning of and the usage of energy. As a matter of fact, the more I learn, the more I am not only confronted with my own ignorance but also, how much more there is yet to learn. This, however, does not involve me in an emotional conflict; instead, there is a calm, dedicated desire to proceed into the countless eons of time which lie ahead in my future in the pursuance and usage of knowledge of Infinity, which is the recreative, interdimensional energy.

The vast and staggering proportions of this great energy—Infinity—cannot be assimilated in a few months or even a few years; in fact, it is in the gradual learning of, and its usage, which gradually begins to supplant the old material drive or libido with a calm, dedicated assurance of this immortal posterity which is inherent and inherited by those who become aware and who begin its pursuance.

Your letter also speaks of acquiring mastership in this lifetime. This, too, is an idle dream and fantasy which is a product of deep-rooted subconscious maladjustments which have created on the surface of your life, a strong escape mechanism. The rest of your letter completely vindicates this statement—your constant indulgence in past recriminations about your sacrifices and sufferings. Your attainments and beliefs in what you think you formerly were are merely devices used in this escape mechanism to try to compromise yourself to various deflationary happenings in your past. This, in turn, has created the supreme effigy of an escape mechanism to become a master, for in such mastership it is, of course, assumed by the escapee that he can now be supremely all-powerful over men and world conditions. This delusion has also been fostered and kept alive either directly or indirectly by various religions. However, it just does not work that way. Mastery over the material worlds is a gradual, long drawn-out process, taking hundreds of thousands of lifetimes lived in these material worlds and the equally gradual assimilation of knowledge which is obtained in the gradual working-out process of the various infinite experiences contained in these lives.

When any person so relives this constant experience sequence and begins to obtain personal mastery over each experience by connecting it with reason and purpose in regard to this experience as being a personal outlet of expression for this great Infinite Intelligence, this then is quite different than the reactionary way which demands a robot-like existence and which, in turn, engenders subconscious escape mechanisms. Almost needless to say, when any person (even as yourself may so do) reaches this far-off position in his future where he has completely emancipated himself from the material worlds, he is what one would call a master. However, in that position, he does not call himself a master; by that time he has worked out his subconscious psychisms which he created in the material worlds. He no longer needs or feels a desire to overcome and master the material worlds which engendered them; instead, he is confronted with a new aspect of his progression. In visualizing the future, he is again confronted with a whole new dimension of unknown and seemingly intangible elements . . . . a new part of this great Infinity. This, again, is a new challenge and in proportion to his knowledge and attainment, he again proceeds to progress into this Infinity.

In our files here at the Center, we have grateful, unsolicited testimonials from many thousands of people who have had wonderful miracles happen in their lives, some of which are even more miraculous than those which happened in the Holy Land during the administration of Jesus—more miraculous in the sense that the entire circumstance was carried on by mail with no personal contact involved. Even more important, these people were given direct knowledge of the how and why, which would, in turn, coupled together with other instructions, enable them to prevent future recurrences or to heal themselves from other future maladjustments.

So let us be practical. If I can give you one single bit of worthwhile advice, it would be simply this: Quit indulging yourself in your past in the negative way, which has, up until this time, been your chief source of personal satisfaction. The past should be looked upon objectively as a series of stepping stones in attaining comparative and constructive knowledge of your evolution—not as a system of self-justifications whereby you can deify yourself. The future should also be so viewed, not with ambitions to obtain mastery over the material worlds but rather, obtaining knowledge to enable you to live in Higher Worlds beyond these material worlds with the complete understanding that you are gradually becoming a more unified, constructive entity in the infinite prospectus.

I might also advise you to stop practicing or using various so-called occult devices such as trying to make the eyes of a person's picture open. Within the dimension of your own mind you can construct whatever you wish. In the realm of occultism, countless thousands of poor souls have constructed for themselves terrible demons which have, in turn, like a Frankenstein monster, obsessed their entire lives. Occultism is only a system of devices invented as a system of escapes, the father of which is that old malformed subconscious.

There is only one way: through the doorway of knowledge and its correct usage, you will gradually progress into a higher state of consciousness, correctly placing your particular position in Infinity, and which is being expressed through your personal day by day experiences. All of the so-called mental sciences, occultisms, mind sciences, religions, etc., which use devices to circumvent or escape the realities of personal existence and position to the Infinite are only expressionary forms of escape mechanisms. These malpractices include prayer, concentration, affirmations, certain kinds of meditation, various other forms of self-hypnosis and are highly destructive to moral and mental integrity. They prohibit what should be a constant, constructive evaluation of experience and your position to the Infinite through this experience.

It is in this particular dimension of understanding and teaching that makes Unarius quite different from any other system of religion or science now presently known to mankind. Even material science does not factually relate man to the infinite prospectus. As of today, mind sciences, like all religions, are still expressionary elements derived from the age-old escape mechanism where, in the inner reaches of each person, he has somehow been made subconsciously aware of this Infinity. So he longs to escape from his materialistic pit of clay and ascend into the sublime reaches of some heavenly place where want and insecurity are nonexistent. In his mad and fanatical desire to escape, reality eludes him, sanity and reason are lost in the tangled thickets of his jumbled life and he is lost until he finds the thread of reason and purpose which will again lead him out of his self-grown jungle and place his feet on the pathway of constructive evolution.

I sincerely believe you will someday attain that position if you continue to search and seek, and above all, learn to discard the falsities of today which are connected to your past.

One last remark: Certain statements of your last letter completely deny and disprove your claim to understand energy or principles of its function. If you did, your life and your letters would be far different. No one could so understand and still cling to the past. Indeed, his very nature is changed in proportion to his knowledge and which, in turn, remanifests as constructive thought and action; nor would he feel the need to indulge in certain psychisms to prove his worth to himself or to his fellowman. However, you have made a great and commendable step in the right direction. This is in the acknowledgment that there was something wrong and you placed this where it should be placed—inside. No one gets anywhere in a self-corrective development program until he can objectify himself as the cause and source of his trouble. Conversely, he can be well and happy. As Jesus said, the Kingdom is within, and there is usually a material hell until the Kingdom can be visualized and proper steps taken in the right direction.

In other words, it's all concept. Within the dimension of your mind you can attain any position you wish. It is wise, however, to have a definite plan of attainment and make constant daily comparisons and analyses as to whether progress is being made or lost. Using your daily life as a criteria, are you constructive and are you productive of right thought and action? "By your fruits ye shall be known."

It is also wise to use science as a basis for beginning or constructive evolution. The world is full of "isms" and "osms". History, past and recent, exposes the falseness of these various systems and beliefs. Obviously, as the Creative Intelligence created all and is all, we can begin to learn about this great "Creative All", through science, and as we learn, we grow.

However, present-day material science is just a beginning; this material science knows little or nothing beyond this third dimension. It does, however, recognize to some degree the possibility of great macrocosms beyond this dimension, and at present, great efforts are being made to probe into this unknown. And here's where you "get the jump" on present-day science; through Unarius you can acquire factual knowledge of the great worlds and dimensions beyond this one. You can also acquire a correct personal psychology which will enable you to start a progressive evolution into these more advanced dimensions.

It's all in the books and lessons. Therefore, I can only urge you to continue to study. Do not become self-satisfied with any future position you may attain; there is always much more to learn and learn to use. Above all, objectify your progress as a development of function with the Infinite Intelligence; not to pacify subversive, subconscious elements which create escape mechanisms that are cleverly disguised and hidden, except in the way they recreate confusion and despair in your daily life.

To your continued constructive progress then, I remain,

Ernest L. Norman

Discussion with a Student

Contrary to common beliefs, the brain is not a thinking instrument; there is nothing in the brain to substantiate the theory that it thinks. It does not. The psychiatrists are very much puzzled as to how it functions, how these thoughts arrive in consciousness. They have what they call an electroencephalograph which is a machine which measures brain waves, but they do not know where the electricity comes from in the brain.

To explain this simply: we have a sine wave, an up and down motion of energy which merely means there is a certain quantity of intelligence which is being scanned from one dimension to another. It is going into consciousness, which is the psychic body which oscillates. The brain becomes an integrator. There are twelve million brain cells in each brain which are like little tiny transistors in the radio. They transform the waves which you call light. We do not actually see things physically because they are transistors, and all that is being done is intercepting sine waves of energy. They go into the retina of your eye and, in turn, pass on into the brain. The brain transforms them into higher frequencies or higher vibrations, which oscillate or go into the psychic body. They set up a ringing situation within the psychic or energy body. They find, through frequency, as they attune themselves (the same as a radio) compatible frequencies which were previously there and which were impinged there in the same way. When this is done, they create harmonics and these harmonics, in turn, oscillate back into the conscious mind. Part of the conscious mind makes the muscles act as a result of this action; the other part causes the voice to activate (or sound) or whatever is necessary to react to the stimulus of what you have seen on the outside. This can be demonstrated in this manner: we can scientifically draw out these diagrams (they have been diagrammed in the Second Lesson Course); we can show the lengths of the wave, what the brain actually does, and where the energy comes from. Energy comes from the inside; it comes from that Father Within—the Kingdom of Heaven Within of which Jesus spoke. That is this great Infinity.

You look outside and see what you call space, but it is not space at all; it is infinitely filled. Because you do not have the necessary higher senses, you cannot see what I can see out there. To me, that space is a tremendously solid sea of energy which is pulsating and radiating simultaneously in all directions, just as it tells in the opening chapters of the Venus book.

Student: My husband who only went to the sixth grade in school is very adept in electronics and has never actually studied it; he says he can never remember the time when he did not know about electricity. How did this come about?

Teacher: Yes, there are many people on the earth today who are from the "old school", so to speak, who came from the old civilizations like Atlantis, where there was a great deal of electronic science existing at that time. They had spaceships and ran their huge cities by cosmic generators and many other such things, so your husband is what we call an old soul. In other words, he learned electronics, to a certain degree, in other civilizations or on other planets. He still retains this memory. As I said a moment ago, all these energies are little wave forms of intelligence that exist within the psychic anatomy, so all he needs to do is to see on the outside some wires. On the sides of those lines or wires oscillate certain frequencies; they oscillate or go back into the psychic body and find other frequencies which are compatible to that electronic apparatus which he is seeing; he oscillates back and forth and there he can do the thing, whatever it is, with his hands.

Student: With him, too, he can do so many things that he has never actually studied or learned, but just has the know-how. He told about atoms, etc., and airplanes long before they were made.

Teacher: Yes, that is very understandable, and I can cite you a specific case which is on file at the War Department in Washington, which happened in the first stages of the manufacturing of the atom bomb. There was a certain part in that mechanism that was necessary to fire it, which was of vital importance to them. The head scientist was very much puzzled; they had striven with the problem for months but could not get the proper mechanism for correct function. All of a sudden, back East, a woman had contacted the scientist and gave him a detailed description of this mechanism on which they had been working so strenuously. She had seen in a dream (the Scientists on the inner had shown her) how to construct it, so he went to work and had the men copy the diagrams—and it worked perfectly! But she had never previously contacted them in any way. It did the important job perfectly. It is described in the Venus book how it is that people are taken in sleep state to Venus and other planets and taught in sleep state many things so necessary, especially for our spiritual development.

The average materialist is, in a sense, like the tail wagging the dog. When we start to see things in the true light, then we realize that we must allow the dog to wag the tail, so to speak, or to let the Superconscious do the directing in all things. The physical is only the surface appearance or the end result of what takes place on the inner. For instance, the building or the floor is not solid; it is only solid to us because we have developed a concept of solidity to it. In other words, we are traveling through space and time at the same rate of vibration as these atoms. If you change the rate of vibration of the atoms in your body, that floor would not seem to exist, or the walls, and you could walk through them, just as Jesus did. This is exactly how He did it—by changing the frequencies. It is a very exacting science.

A Student Letter on Maintaining
the Positive Versus Negative

The following is part of a letter dictated by the Moderator to a student—the one who in the past was old King Herod I. Such concepts can serve to help any person, thus the inclusion:

The most important problem confronting the aspiring student of Unarius is self-mastery, and it is almost needless to point out that until certain definite steps and accomplishments in this self-mastery have been accomplished the aspirant has no chance of becoming a better person.

The whole context of the Unariun concept is based upon this premise: you cannot become a better person until you have mastered your personal emotions. You cannot hope to escape the purgatory of karma which you have heaped upon yourself in your past lives, no more than you can hope to escape present and future inflictions by indulging yourself in emotionalisms. Therefore, if you find yourself at any time, now or in the future, emotionally involved with any person or persons or any set of circumstances, the first step in self-mastery is to realize that you—and you alone—are to blame and are responsible for your involvement and for your emotion.

When you have accomplished such real, honest judgment against yourself and you can judge your position on the basis of all factors involved in the situation, then you shall find that these emotionalisms will be discharged; they will lose their power to hold you and your future will become freer from recurring situations.

Self-mastery does not mean the suppression of emotion; it means discharging and discarding this unintelligent reaction by a sequence of logical interpretations which are taught throughout the Unarius liturgies and are the basic elements of personal psychology. When you properly understand human nature, without involving your personal ego in this understanding, you will be able to quickly settle all differences which may arise in your mind on the basis of logic, reason and understanding, and when you have learned to understand yourself and others, as well as the world about you, you will be able to automatically project a strong, positive vibration or power into everything. People and dogs will like you without the necessity—as you now think—of trying to impress them with your ability and your desires.

This strong, positive vibration should not be misconstrued; it does not come from an egotistical person. Such persons only incur enmity and dislike. The positive love vibration comes only when you understand yourself and others and you are interested in them above your own welfare and regardless of personal benefits.

Therefore, to anyone who would like to live a better life, who would like to serve mankind and the Infinite Creator, he must first accomplish this first and all-important step—self-mastery. It is a long step and perhaps the biggest step you'll ever take, and when you have accomplished self-mastery you will no longer find yourself emotionally involved with others. It will not be a question of being welcomed into their homes or in their company, nor will there be any question of your abilities. All such things will be automatically adjusted and taken care of beforehand. That is part of the principle of Infinite Regeneration. As you become wiser and more understanding, so does the emotional world pass from you in proportion to your understanding. The degree of your emotional involvement then will always be a personal yardstick whereby you can constantly measure your success or failure.

If you find yourself in an emotional situation, just say to yourself, "There is no one else to blame but myself; I have let myself become emotionally involved and there's only one person who can change it—that's me," or words along that line. When you can honestly and objectively analyze, within yourself, that emotionalism is a form of mental weakness, that emotionalism is indulged in by people in lieu of an intelligent, understandable way of life, then you will be able to quickly change your situation.

Remember the old classic quotation, "Evil is always in the eye of the beholder". An evil circumstance only begets its own evil progeny, et cetera. Don't blame others—blame yourself for being ignorant and failing to understand either yourself or them.

Moreover, in your quest of self-mastery, so long as you are living a mortal life and until you have evolved to a higher plane of consciousness, you will be to some degree, victimized by the emotional world. You must not, however, attempt to console yourself by saying that circumstances were thrust upon you or that others do it, so why can't I, etc.

Living in the material world means constant involvement in emotionalisms. You will, therefore, be constantly called upon to exercise the franchise of intelligent, logical analysis. You will also be constantly called upon to dissolve these negative energies by reforming from them constructive thought and putting this constructive thought into action by understanding other people, seeing or anticipating their unfulfilled need of such constructive thought, and by objectifying your own need and lack, as well as that of the others, fill these voids with this constructive thought which, when properly tuned to negative situations, will reverse and cancel this negation. These are simple elemental and scientific astrodynamic principles used by the Infinite Creator in creating Himself in all things.

If you wish to become a constructive, creative individual, you must learn to use these principles. If you are negative and emotional, you will become increasingly so as the years and the lifetimes pass. There's only one way to reverse degeneration and obliteration; through constructive self-development, learning to understand yourself as an evolvement through an emotional world. Your present emotional position must be changed to a constructive attitude which uses the positive power of intelligence, logic and understanding.

Learn to be interested in others; be compassionately minded with their problems. Don't toss your ego around; if you are something of any value, people will quickly recognize it—you won't have to tell them. If your life is empty and lax, then fill it by helping to fill the lives of others—not with your self-importance but rather, with love and understanding. Never try to inflict yourself upon others; instead, create of yourself a storehouse, then they will come to you.

Self-mastery does at once accomplish all the best and creative desires and fulfills the promise of Infinite Creation.

The Unariun life is lived in peace and complacency; it is lived in harmony with all others and all things. It is lived without turmoil and strife and constantly is filled and refilled with an abundance of all virtues.

The Moderator's Reply to a Student's Questions

Student: A lady friend often "sees" layers or clouds of luminous colored gases in the atmosphere. Any significance to this?

Teacher: I can tell her what it is. It is simply an ionized layer or mixture of gases that's put there. Man made it.

S - What kind?

T - Gases in another spectrum that is not known by the earth world scientists; xenon, freon, krypton, or some of these other rare gases of the earth. We can suspend an ionized cloud of gas. Now a cloud is an ionized cloud of vapor held there in molecular suspension because we have a minus to plus one ratio of ionization. That's why it is held there.

S - Doesn't the spectrograph relate to these things?

T - The spectrograph is a crude thing. For instance, the scientists are all mixed up on the red shift. They don't know what a red shift is. According to the red shift, everything in the universe is going away from the earth at almost the speed of light—the Doppler effect —the same as the whistle on a locomotive coming toward you or going away. Well, when they take the spectroscopic examination of the elements of the light from these planets or these stars and some of them are so bright they don't know why they appear to have hundreds of times more energy than is, to them, physically possible; they can't calculate the energy so they try to place these different bars of these different elements in their proper place according to the spectroscope or to spectroscopic examination, where they should be. Everything comes out minus in the graph and the red is way over there!

So what is it? They don't know and they can't tell you. It's very simple if you can analyze it on the basis of what problems we've had in hysteresis and magnetic lines of structure and how things come into this third dimension in its time-phase relationship. They figure it out simply as a question of distance. It is in an equivocal situation where we say we've got millions of light years between here and an object, and it it's going to take red a little longer than the other rays to count. It's not that at all because all of these things have to be figured out in an entirely different way, or different basis.

S - What about the blue, the ultra-violet spectrum? The understanding that the universe is a very large size is detected on the blue shift.

T - It's much the same proposition as I explained a moment ago; it's impinged emanation of a tremendous amount of energy. They can't calculate with their physics; it's beyond psi with them—what they call psi.

S - Is this a very large universe?

T - Not necessarily, because we can be concerned as we did before, with the vortex or the centrifuge of energy. What we would have to figure is the ratio of hysteresis in the third dimension, because therefore we come a cropper between time and space.

S - Could it be a vortex?

T - Any of those bright, pleasant things out there you call suns are supported by a vortex; they're the nucleus of a vortex. They have a part you can't see because it's all radiation from another dimension. It's all energy wave forms—a centrifugal proposition, a spiral nebulae proposition, much the same as our sun or our universe, for that matter. The pattern is universal. Now what they can see are different cloud forms, the nebulous masses out there. Well, they just relate merely to ways in which they are seeing energy in the third dimension. There are other different types of energy sources from the fourth dimension, but here again, they seem to think they exist there per se. They've got this fixed proposition. They don't realize it's only a secondary radiation or emanation from another dimension. Yes, the key is right there. Everything is predicated in this dimension. Obviously they're going around in circles. I think someone figured it out at one time that there exists more than three hundred sixty billion horsepower an hour in power on the Sahara Desert alone! Now where does all this tremendous radiation come from—this heat, this conversion? When you figure the whole of Infinity and then figure the earth is only a tiny little speck in the receiving, think of how much radiating power goes out from the sun every hour! Where does all this energy come from? The scientist thinks it's a small thing and that the atomic energy which could be converted from the sun wouldn't last a year; besides, it couldn't be controlled. There's only one way it can be controlled and that is under the regular laws of the cosmic interdimensional energy transmission, as I have described it in my books—at least to a degree in which people can begin to understand it; that is—the formation of it.

Now if they want to go a little farther than that, they can estimate wave lengths and other different measurements. We can do that too, but they will have to understand it in another dimension in which the present physics and mathematics now can't compensate or embrace. If we are going to understand the fourth dimension, we will need an entirely different mathematics.

S - We must first master what we have now—what you've given, must we not?

T - Well, this is exactly what I said awhile ago. If I leave them alone, they will get it in time; at least, some of them will. And the others that don't, well, they'll have to come back down into these pits of hell and get their heads pointed in the right direction again—even as I'm doing now. That's our problem; we're sort of a life-guard of the fourth dimensional cosmos—the lifeguard service always going in to serve and rescue a few of them.

S - In another vein, what did Jesus mean when he spoke of being in the pits of hell?

T - You mean during the time he was crucified and resurrected, "I've been in the pits of hell, teaching the lost souls; some of them have been there, yea, unto the time of Noah." Now that's a Biblical translation, but actually He was as much in the pits of hell, which is right here on earth, as He was in any other place. He was connected with people who have used the earth per se and considered it the only and all that there is. But the lower astral worlds are also filled with count-less lost souls—in hell, so to speak.

*   *   *   *   *

Within the scale of evolution, it can be sufficiently equalized that with a certain set of basic values, any human can live or be expected to live a reasonably equitable life. It is not presupposed that these basic values could or should include any of the old super-stitions of the past. It is known and recognized among the young people, the coming generations of this day and time, that there is a beginning of the discontinu-ance in these beliefs, that in the sophistication of this age and in their generation, these people do not be-lieve in the superstitions of their elders. Unfortunately, these growing generations still do not have a philos-ophy of life wherein they can find realistic basic values which will relate them to creation and to their moral responsibilities to themselves and to this creation.

Instinctively or generically, the birds and the beasts of the field each relay their own specific relationship to creation. Each one manifests and gleans, according to this position on the scale of evolution; yet, because man is, individually speaking, a more intellectual person, capable of relating many factors and facets which are otherwise unrelated, he has attained a certain scale of mental evolution which will either advance him beyond the domain of this physical world, or it will destroy him and plunge him backward and downward into the hellish labyrinth of moral and spiritual disintegration. That this latter situation is already occurring and has occurred for many people is manifest today in even a greater degree than any time in history, and it will continue to multiply so long as basic values are lacking.

It would be incidental at this time to mention that there are gleams of hope and light on the horizon of our society. It is well known that our present church hierarchies are losing their power and their domain is passing, a fact well recognized by the archdiocese of the great mother-father church. The Pope, himself, has instigated and helped to pass certain church legislation which has changed, modified or otherwise adulterated time-honored and hallowed concepts and laws by which Catholics have lived for nearly 2,000 years, in areas of diet, birth control and in other areas also. There has been great liberalization in an attempt to waylay or to avert what has been clearly prophesied and written upon the wall by the hand of time, that within the next hundred years or so one would see the complete dissolution of all Christian churches, or at least, a complete change in the administration and in the tenures and concepts, and the expressionary elements of these churches.

(This letter was written by the Moderator to his sister in Utah, who was unfamiliar with the Unarius Science and should be most interesting to any student.)

August 15, 1962

Perhaps I should forewarn you: that which you might read in the books may shock you; however, always keep in mind that it can all be proven scientifically and historically. Every day science is proving what I have written even many years ago. What I am teaching is, in effect and in reality, the "Second Coming"—a big statement, perhaps, but think for a moment: Do you think Jesus would return as He was 2,000 years ago? Indeed not! He knows that the world would destroy Him again in some way, manner or form, if He so appeared, nor would the earth be burned first. That's nonsense; you can't destroy evil by fire, it has to be replaced with good—in other words, understanding Creation.

So if Jesus "returned", he would do so in a very subtle way—through the mind of a person like myself who understands Creation. The Message of Jesus would be the science of life as it is lived here—and especially, hereafter. This Message would come to those who had prepared themselves for the advent— by preparation in Spiritual Worlds where they had lived many times in between earth lives. Then as they "lived" again on earth, they could join heaven and earth in the great reality of Creation.

This is the Message in my books—a Message which must become the reality of Creation before any human can live hereafter. First, the earth or material world must be gradually torn down (in consciousness) and replaced by the true science of Creation. The material world is a world of symbology; all things that you do—your thinking and acting—are motivated as a reflex from these symbolic forms. Even your religion is a symbol.

Now symbols are false; that is, as long as there is a symbol there is no need or necessity to think. Thinking, therefore, is not thinking but is a reflex action. You think only when you are able to see the entire creativity, the plan, the evolution, etc., which created all things. Symbology is only a crutch, only a device used by elemental minds until they develop the intellect to a point where they can see the entire creativity in all things. Jesus called this, finding the Kingdom Within.

My works are therefore a logical, scientific approach to thinking, to replace the symbolic world with the whole of Infinite Creation. This will be your "savior". When you can so begin to conceive, you will be "saved"; you will begin to find the Kingdom and eternal life in the finding. The "Father" is part of this. The "Father" is the Creative Principle of life which creates and re-creates forever in all manners and forms, atoms, worlds, universes and heavens, and if wrongly understood and used—will be the hells.

Any person's fears, insecurities and drives, etc., each is an anxiety to realize more of life and of creation before he leaves this earth in this present life and returns to some spirit world. When you came into this world you had a plan. You said, "When I get back in the physical world, I'll do better," etc. Now, you subconsciously know all has not been attained, thus the urgency. It is also insecurity.

When you read my writings (the Unariun liturgies) you will realize much of this is still unattained; you will also be prepared for even greater attainments in future lives lived on this or some other similar world. It is not what we do in a physical way that counts; it's what we learn that's important. Learning of Creation is learning to live eternally in a very much more advanced state. Life is actually only consciousness. As we conceive, so we are (what we conceive). We should therefore never place any limitations; never say "it ain't so", etc., but rather, try to understand, and as we understand, we grow. Knowledge of the Infinite Creator gives us power over life and death; it gives us mastery over all things in proportion to our knowledge. Through my works, you are gaining this knowledge and which you will later begin to use in reconstructing a new life in a higher world. In this way, then, Jesus has returned to you and you have been "saved" by knowledge and usage.

We have countless hand-written testimonials from many thousands of students who have had great miracles happen to them; terminal cases, mental and physical conditions of all kinds. There are at least fifty thousand or more people who have been helped— many not even realizing it. They are people from all walks of life, all professions; in fact, every person I have ever been near or spoken to has had something important and wonderful happen to him—even though he may never know. It was that way when Jesus was on Earth. People who understand Creation are, in effect, a great powerhouse; intelligent energy is pouring from them in all directions into all things. This is a simple statement of scientific fact. I'm not boasting, and if I am, it's to impress you with the Creative Principle in action—which religionists call Christ.

Christ is not a person; it is this Creative Principle. In my books it is first described as a scientific form, a sine wave. Later on it is developed into Creation or the power behind Creation. The Father and Christ are synonymous; the Father is Infinity; the Christ is personal realization of the Father which again recreates for us a "heaven" or a state of consciousness which is complete —no fears, frustrations, etc., just peace of mind and the great power of understanding.

One more point: as you begin to understand, then you can recreate your relations, friends, etc. This is the way to "turn the other cheek". By understanding our fellowman, we can project to him a bit of help even though he has slapped us. We must first, however, understand ourselves before attempting to help others. And that's all in the books.

As you no doubt would assume, all this study will require great effort, not only in reading but in application and by seeing Principle in action all around you. Try to conceive, objectively of course, as much as possible. This will be the way you find your answer, or more abstractly, we never really attain the complete answer. That is the challenge of Infinity—and even Infinity presents the same challenge to Itself, for there is no point of termination.

(The following was a greeting sent out at Christmas time to Unarius students.)

We, the Unariuns, are mindful and aware of the approaching Christmas season and a day wherein the Christian earth man observes the birthday of the Nazarene. We are also mindful and aware of many other things which are attendant in this observance.

Like many other seasons, this one, too, will be observed in great hypocrisy. While all things are being done, either joyously or by necessity, even in general conformity, there hangs behind all this the heavy pall of fear. The clouds of war, dissension and strife among nations and peoples hang heavily on the horizon.

We also know that these same people will, in a short time, be observing the crucifixion of this same Galilean —a custom strangely mixed with new life and the belief that death will also give life to those who believe.

And again it is, that as these symbologies are great hypocrisies, just as are all other symbologies, the greatest of all hypocrisies is in the believing, for as a man believes in a symbology, so he walks in its shadow— a shadow of the blackest night, not penetrated by the Light of Truth.

If you would observe the birthday of a Savior, then let it be that day when you shall step aside from under the shadow of symbology, for this will indeed be the birthday of your salvation. Yet, if you would have this salvation, then surely you must prepare yourself for it —not in blind faith or in the belief—save that these things are joined in the acquisition of knowledge of your future life, for surely as knowledge is gained, so belief is strengthened, and ceases to be a symbol.

Christmas then for you, dear student, should not be the garish pagan symbology, lived in fear and shaped in conformities, rather, your personal birthday—a day to begin all days, even unto Eternity—each day lighted by Truth and bringing with it the knowledge of Immortal Life.

Note from the Moderator 5-23-65

To a Visiting Student: Ruth is one of these manifestations of which I have spoken—the completeness of the Absolute and how you understand it because for many years I described Ruth to people all over Los Angeles many years before I ever knew her. I told hundreds of people I was looking for her and what she looked like. She was the only person who could understand my science and had been preconditioned for it at that time because there are very few people who could live with me on that basis, not because I'm irascible or temperamental or anything like that. I'm fairly average and normal in my reactions but simply when one is transcended and lifted when these scientists and the Great Ones come in—and the way in which they come in, it is beyond the average person's comprehension because it is done so naturally. It is done without fanfare and trumpets and all this kind of fuss in which people have formerly associated the appearance of the Great Ones but it is a simple, living process. It should be well understood and compatible with everyone's understanding and it will be eventually.

October 9, 1955, 11 P.M., in answer to our disappointment of the publishers who were to publish our first book and we found them to be dishonest, etc.:

Dear Ones, be not sorely troubled nor heavily laden, nor let the weakness of other men's minds become burdensome obstructions, for truly, each man plays a small or a large part in history, according to how he so determines it. We who have ascended to a higher plane of concept can view more objectively the histories of the many worlds of which your planet is a number, nor are we concerned with the destinies of man collectively or individually in such a manner as it would obstruct his personal evolution. We do not wish to be misconstrued that we are not sympathetic or understanding, for wisdom can be administered only in the light of pure wisdom. For surely, if man gains not his own experience of life, how then can he be master of life?

So it is written in the histories of your world, and so it is thus in such planes as these and in such proper time that man springs forth a creature of carnal lusts and desires, and from the jungles of materialism he must learn to thatch the roof of his new spiritual dwelling. Down the many corridors of time move the legends and hosts of shadowy figures who have once left their footprints in the sands of time upon your planet, and so they have come and gone and come again upon a newborn day.

Man on your earth plane today is, as he has always been in such an environment and in such a dimension, a creature motivated by the same animal instincts as the beasts of the fields and forests, and the fowls of the air upon which he feeds, and yet with all this, there is somewhere within, a distinguishing difference that links him with the great creative force of the Universal Mind.

At present you live in a world and a time which is unique and unparalleled in the past history of your planet. Many civilizations rose and fell, some large, some small, and yet, none with the largeness of the great masses of population which surge and beat upon the surface like the ceaseless waves of the ocean. And so the time has come when each man lives upon the other, and while he eats not of his flesh but only of those things which are about him, his properties have become chattels and he knows not the day when all that shall be scattered will be as leaves in an autumn wind.

So it is with you, like the parable of the Tower of Babel, that as each man who dares to raise a spiritual edifice into the heavens of his Father, if he has not great strength and courage which is given him, surely he must be torn asunder; nor will one stone be left upon another, nor canst the nails or bindings of sinew keep the beams together, for the screaming winds of ten thousand discordant voices will be raised against thee.

Each man comes into your world at his time and place and selects from it those things which are necessary for his growth and development; nor can it be said that he is like the corn of the field and that the corn would say to the barley: "Wherefore am I not barley, for surely he is much more graceful than I?" Nor canst the barley say: "Wherefore am I not corn, for is he not treasured in the granaries of man, while I am fed to the asses of the field and the fowl of the barnyard?" Yet, surely it is not whether he is corn or barley, but each man should taketh his place among his fellowman and become all of the things he is.

So fear not the voices that shall be raised against you, that they may seem like the rushing winds of heaven or as the roar of the lion; nor shall they come upon you while you sleep, for the wombs of time have begot many such worlds as yours, and the races of men are more than the grains of sand, yea, even more than all the worlds.

— Enoch

Would You Have Immortality—Immortal Life?

Unarius and the proposition of Unarius is, in its totality, entirely different from any religion, theosophy, philosophy or science as it is now concurrently being expressed. The existence of such creeds, credos, philosophies, etc., are in themselves, principally palliatives or placebos developed and exercised in common usage as a pressure valve used to relieve pressures or to otherwise justify differences in individual lives; and all such palliatives and placebos are therefore subjective to individual interpretations.

Science too is, in its factual differences, guilty of the same misinterpretations and justifications, false importances, etc. Unarius as an interdimensional science first presents the interdimensional cosmos and immutable laws which govern and control all functions within this infinity. Moreover, these laws and all creative principles involved are not subjective to individual interpretation or even categorical classification. Indeed, these laws and principles are not, at this present time, known to even the more highly-developed scientific elements. The proposition therefore remains inviolate.

If immortality is to be attained and if an immortal life is to be lived as a progressive function of evolution, then it becomes the individual proposition of every human to learn of these creative laws and principles, their functions, etc., and to integrate them as functional denominators in his daily life. Furthermore, this proposition clearly defines such immortality; and immortal life is lived in dimensions or realms not physical or third-dimensional in nature. Therefore, again the importance of learning interdimensional law and principle becomes life-saving as life in such higher interdimensional realms cannot be lived in a third-dimensional physical body or with a mind which functions or oscillates with third-dimensional constituents.

Therefore, to achieve such immortality and an immortal life, there must begin an inclinational process of knowledge and wisdom which explains the interdimensional cosmos and life as it is lived in these higher realms. Through the various books, texts and other liturgies of Unarius, this vital knowledge is immediately accessible to any and all humans who have the necessary requirements which will make Unarius desirable and comprehensible. The individual must first have passed through the realms of common atheisms which have commonly been used throughout man's written history. Such atheisms as religions, philosophies, sciences, etc., have all universally subverted a true interdimensional prospectus as it has always existed.

Religions invariably use some irascible temperamental godhead which functions on an emotional level. Philosophies only point out vague differences in individual interpolations whereby opinions can be centralized as an individual escape mechanism. Science is embroiled in an inescapable atheism which relegates life and all origins as third-dimensional.

To escape all these pitfalls requires more than herculean effort as each individual life started or starts in its third dimensional evolution from primitive beginnings. Structurally speaking, such lives are compounded wholly from third-dimensional life experiences which have been many thousands of years in the making. It is not possible, therefore, under these circumstances for any person to summarily read through the Unarius liturgies and be completely and compatibly developed for life in a Higher World.

Again the proposition clearly defines that interdimensional knowledge and its insemination become an evolutionary process—a complete totality of personal involvement in this process of insemination—a process which can be expected to go on for not just so many years or the remainder of a lifetime, but a future of many lifetimes lived in a third-dimensional world and in lives in between lives where schooling becomes particularly advantageous. In other words, in the millennia of time ahead of this person, there is a reconstructive process of the psychic anatomy, the development of an energy body which can live in a Higher World. To return to or reincarnate into physical worlds means a constant and an ever-increasing comparative equation between physical life and life lived in a Higher World. Comparisons will involve, in totality, the third-dimensional physical world and the interdimensional cosmos.

The importance of maintaining contact with Unarius has now been multiplied many times. It cannot be a simple rote process, a simple reading of the books or texts; it must be an all-consuming objective which totally involves the individual. As books and texts are read and reread, newer and more profound truths are found as the individual intellectual level expands and develops.

Equally important is the constant contact mentally made with those highly-developed Beings who, for convenience sake, are called the Unariun Brotherhood but who, individually speaking, would be in biblical parlance called Archangels. So you see, dear ones, the importance of keeping yourself "plugged in", as it were, to the universal, interdimensional powerhouse of wisdom—the interdimensional infinity—the source and structure of all worlds and dimensions. To learn and understand all this is far beyond the capacity of any third-dimensional mortal, yet a beginning must be made, a preface to a constructive evolution and its clearly marked pathway into the Higher Worlds.

Read then, and study; learn to evaluate your environment, your life, as a function of law, order and principle being constantly expressed and re-created in a countless multiplicity of forms and motions—all originating from and controlled by interdimensional law and principle.

Philosophy by Lahamahasa

1959

Greetings, dear Sister. I came in on the frequency when little sister stepped out. The name you were searching for (Lahamahasa) in the book was mispronounced; however, that is customary sometimes to mispronounce names, especially through the countless centuries. Many times the races of people on the earth have personified all the various types of evil and good in some person, when they find a person who symbolizes the good or the evil. It is a poor man who does not live up to his expectations. Vishnu is the personification of mental attitudes in the progressive structures of man's advancement. Vishnu incorporates idea and form and consciousness of progressiveness of man's spiritual nature. To capitalize this in the individual sense and to personify an individual has always been a common form of ideology among the people of the earth. Man lives from one hour to the next as far as his earth life is concerned, believing in only the things which he sees before him, not realizing the vast and hidden potentials which would liberate him from the karmic life.

You are troubled with concept, perhaps I can help. It was expressed thus; that concept in the beginnings of the more abstract consciousness becomes not only the motivating force of your life but also becomes your personality. It is so that no one can lie by the side of the road of life for an indefinite period, but must always move with the stream or the tide in one direction or another; as he goes, he becomes. All things are added unto him as he sees and believes. Concept is the living essence of creation in all forms of consciousness. It has been given various different translations, but always does concept remain as a singular pronoun in man's consciousness. So as he is and becomes in his earth life, he worships only one god at a time and in only one temple before he goes on to the next, believing that in each observance he finds a token of kinship in the God Force in which he believes. And whether these things are found in temples or shrines or whether they become embodied in different ways of his life, always they are things which he has taken unto himself—sometimes more fondly than his life and loves them more. Always he passes from one temple or shrine to the next to renew his acquaintance with a new god or a new form.

There is much to be given, so much to be brought into consciousness if you have the patience, the forbearance to surmount the difficulties and the obstacles which will incarnate into your pathway. It is a law of karmic structure that no good can be expressed without diametrically opposed differences of negation. The trick here is to find ways of discharging or circumventing the evil or negative force without involving personal destructiveness.

One of the simpler devices of the earth man is the escape mechanism, sometimes found in the barrooms or cinema palaces. Perhaps escape mechanisms are found in other ways, too, but in your particular case with Ernest, escape mechanisms should by no means include personalities or differences of opinion between yourselves. Seek to find an outlet for the preponderance of negative forces which is the ultimate end of the other cycle of positiveness expressed in another direction.

As the Avatar Jesus said, "As ye seek, so shall ye find." There is much psychic destruction, all of which occurs in moments of stress of that nature—like the proverbial one step forward and two steps backward.

The fallen man is only defeated if he refuses to rise. The thought or idea, form or consciousness for the individual can be twofold in purpose—it not only becomes something which lives and recreates itself in your own consciousness, but also, it recreates itself in other people.

The idea here, of course, is to recreate consciousness in a constructive form. This is the basic law of progressive spiritual evolution of which we must always be conscious. Mental discipline consists in constantly ejecting negation, refusing to be a party to it, always supplanting negative thoughts with positive constructiveness, for in that way we then begin to rebuild the spiritual body which you call the psychic from constructively activated spiritual elements of idea, form and consciousness which relink you to the Infinite Cosmogony. There is no other way to attain these things, contrary to some of the preachments upon the earth today. We have all found this common road together and pass into the future.

The only terminating point of consciousness is attained when the person believes he has reached that terminating point by all things of consciousness. They become nonexistent in their present form when they are viewed from a different angle. There is nothing about you in your present life which tomorrow you will not view differently. Again these things have reconstructed themselves into new idea, form and consciousness, for such is the way of movement of intelligence, consciousness in this infinite world. As it has been expressed by Ernest many times, consciousness of all this motion is contained in the opposite of polarity which again brings us back to that very fatigued point of consciousness in your own personal relationships to each other. You see, dear Sister, psychosomatic science can be extended into many different ways of consciousness. Your escape mechanism should be one in which you can direct more of that constructive Light from Infinity upon the negative terminus of consciousness.

I can assure you that there is nothing more poisonous to one's own nature than something which he has created for himself, for as it is a part of him, it lives and breathes the very essences of his own creative effort. It is much worse to remove than any astral entity which he may have attracted by negation—and a common pitfall indulged in by so many countless millions. We must learn to activate into positive consciousness the negative portions of these expressions which creep into our lives. The further you advance upon this pathway, the more necessary these things become to you if you are found in these primitive earth-plane existences.

The fruit in the Garden of Eden on your tree of life has a wonderful and beautiful appearance, but if taken internally is very poisonous in nature. That is the true interpretation, spiritually speaking, among the many translations of a parable of such universal nature. Yes, in India many thousands of years before the advent of Jesus, we too had a parable of the Garden of Eden tree of life. It is written in the temples upon the stones and blocks which make the temples of India, to Vishnu (the preserver of life), to Brahma, (the Father of all and the Creator of the Infinite Cosmogony). From the most finite atomic forms to the most universal consciousness in Infinity is carried the great idea of Brahma, but to Vishnu becomes the individual form, the superconsciousness, the preserver.

A Message to Ruth—December 31, 1959

If you wish to turn on the little box (recorder), you may do so. I shall try to overcome whatever passive resistance may be in the subconsciousness of Ernest to bring you my own personal message for the New Year. As you no doubt know by studying various historical configurations of religious and other documentary evidence which is contained in the archives of the historical world, and as it concerns man's spiritual progress into the Infinite, and as Ernest is so scientifically explaining his teaching as is contained in the expression from great cycles as manifesting as Spiritual Consciousness, always man in his search to orient himself into a happier and more compatible state of union with the Infinite, seeks out and brings into his daily life, various and different configurations which have been termed as religious or cultistic observances.

And so the histories of the world have been filled with numerous depictions of pageantries, fiestas and various other ritualistic observances which have their roots deep into the very subconscious fiber of each human being and in his appeal to what he feels is some vague internal connection to the Infinite.

The mystical and auric forces which move around in great and preponderant abundance have always filled man with great awe, with a great sense of mysti cism, and with a failure to understand. And in these things he has only tried to depict in his own way, what he believes should be the right thing to tender in his emotional reflexes to the great Infinite, as he divides it into his consciousness, whether it resides in a superficial form of consciousness of some God, as in a Christian, or whether he adapts within his consciousness such configurations which involve one hundred or even one thousand different deifications, all of which have their own particular and specific point of equation in the emotional values of man's life upon the planet earth. And so man has lived and lived again and begat into each generation and unto the progeny of the following generations all of these things and many more.

And thus even the seasons of the year have been interwoven in this fabrication of life, and strange, indeed, that while all of the content of man's life is lived in the mental stance and in spiritual configurations which involve unknown deistic forces which are beyond his comprehension and understanding, yet he refuses to acknowledge these dominant and spiritual influences into the reactions of his own personal interpretations to his fellowman. He has always maintained his war-like attitudes, his lusts, and the various tenacious and strenuous ways to wring the very essence of his life from the bowels of his mother earth. And so the seasons rise and fall, and with them bring new interpretations, new configurations of some ruling deistic force which is so contrived in his own imagination that it must be the leaving off of some particular form, and with another, it means the resumption to again bring into the succeeding days, weeks and months, some new and different set of interpretations.

But how well it is put in your works, that man always turns his face to Infinity, and in the day-by-day turning, finds that there is a new interpretation. So man is never, in himself, static, nor are any of the other forces of the world he lives in likewise static, but all maintain the same basic movement into the Infinite. Even the atoms in your body are undergoing the same changes, and they in their own configurations of molecules and cell structures are being constantly replaced because the atoms, themselves, represent only a form of consciousness of the Infinite.

80

And so into the future, man will learn and he will find the greatest of all messages which will be interpreted into his daily life, and that it is not the conflux of any one or a great number of seemingly external or mystical forces which reside in unknown dimensions about him, but that it is the way in which he interprets his own life from within his own mind. In this interpretation does he incept into his own consciousness certain relative qualities of the Infinite Light which will develop and change the seemingly drab and animal-like form of the man who is coming up from out of the more primitive developments of consciousness into something which is indeed possessed of certain potentials which are as deistic as those which he has so contrived within his mind.

And so my message to you, dear one, this eve as the New Year represents to some degree, calendrically speaking, a turn or you are very vaguely conscious that it is in the presuppositions of the subconscious mind —vaguely reactionary in its pure sense and essence— that you must, in some degree, be mindful of the past, present and future in your small world of which the physical finds you. But let not the false appearances of this physical world confound or confuse you. You are indeed living in the pure spiritual in every manner and form, and indeed, all that of which you have spoken, you recognize in your own great metamorphoses that it is indeed true you have turned your face to the Infinite, and in the manner and form that great Light has been shed into your consciousness, this great Light has changed you, so far as you are conscious, with much more than you could have voluntarily done in many thousands of years treading the same reactionary trail of existence which is the lot of most mortals. Yet there are times and sometimes there are places when there seemingly comes into one's life a great helping hand— a hand which has been prepared and strengthened with ages of wisdom—and with the living of one's self on the inner and in conjunction with the higher denominators of spiritual interpretation, this great hand can help if one can reach out and touch this hand; he can be lifted.

In all of these things, they must be made equal in the eye of the Infinite, for such things are not made possible without great preparations, without much setting aside, without a high degree of realization from the inner self. And while you externally consider yourself fortunate, yet this is, in itself, one more of the phases of your personal realization of internal and spiritual consciousness.

My dear Sister, you have earned what you are now realizing to a great degree, and whatever manifestations comes to you is a direct polarization with a true understanding of love, and in that polarization of love from someone who understands the facts of life as they should be understood, then under these conditions all things are equal in the eye of the Creator and you do, indeed, become a son of God.

And so my message to you both is one simply this— that the New Year and each succeeding year will bring forth its own abundance and its own realizations in direct proportion as to how well you live with the consciousness which you have, and which comes from within, and that you can set aside permanently and forever small and personal differences in your physical life, for these things only become the way in which you could destroy yourself. And if this is remembered, have no fear; your journey into the future will be a continual revelation of new truths. It will be brought to you in abundance in new realizations and there will be evidential manifestations in whatever you touch in the world about you, that your feet are, indeed, well placed upon that spiraling pathway into the stars.

Paramahansa Yoganandaji

If thou wouldst come to me, my Love
  across the span of distant shores
My heart would seek thine own
  to beat in harmony
To lift the souls of man
  where e're that he may be
To heights as yet unknown
  untrod by such as he.

For Heaven but awaits his reaching hand
  to bid him enter in
For all the beauty and all the love
  the great Creator has for man
  is reaching out to all
He knows no limitations nor brevity
  in all his gifts to mankind
But to one and all He extends his all -
  the infinite nature of Infinity.

Uriel

Sufficient Unto Each Day
The Evil Thereof

1955 —

Personal conversation to Student from P. Quimby: "I know how you stress personal identification; this is why I am repeating my name: Phineas P. Quimby." (He was the teacher and instructor of Mary Baker Eddy— and a very learned soul in the understanding of man, spirit, etc.)

My dear Sister, I have been working with Ernest many years, and no doubt you have often felt the reflection of some of the Christian science dogma in his teachings or explanations as they were contained in the book by Mary. You remember these things, do you not? I have noticed that you have been troubled with a recent exorcism which took place with you, and you must remember first, that the point in mind was at that time, the lower self as it concerned previous life-times and dispensations which were negative in nature. As you are, too, a metaphysician of some understanding, you must remember the first essential of good metaphysical practice, which is, that you should not lend power or strength to negative patterns by acknowledgment, by constantly repeating them, for this, in turn, gives them form and an affiliation with the future which is equally as destructive as they have been in the past.

In more technical language, we can refer to the pure abstractions as concerning movement of thought form or consciousness from positive to negative cycles of movement. And this, too, has been very well explained to you in the past several years, and I need not repeat them. I have only one question to ask; what would happen if you did the same thing to all of the things which you personally know about, if you would give the same credence to the negative end of the cycle? Everything would soon vanish in mid-air, would it not? As the great Master Jesus once said, "Sufficient unto each day the evil thereof," and this is exactly what He meant —that we all gain a definite perspective of vision, a new horizon, a new comparison in our present position on our scale of evolution, by comparing our present with our past dispensations.

If such things have been even partially destructive or displayed such idiosyncrasies of nature which could be considered anything but a true metaphysical bent, then there was indeed time for change, and perhaps change should have taken place even before. So how else can these things in their negative nature be considered anything else except by the mere virtue that they give us perspective; they give us a point of balance or equilibrium whereby we can motivate future activations in our daily life. Is this not so?

For your sake and those who love you, and those who are reaching out to you, it is very important in this crucial period of time that every effort be extended in a positive way and that a strong conscious abstinence from any expressions which could be considered negative in nature, or to thinking back into past dispensations where they were of such.

I would say yet one more thing; that you have not been chosen for this work; you worked your way to it. You occupy your position in walking with the Masters by virtue of these efforts. Only one thing remains in your path—the past negations; and they would indeed be fatal to you if you repeat any of them in the future, and especially in the near future. Even credence of thought into such directions as have been displayed and which were completely and thoroughly exorcized from you and these things cannot return to you unless you so will them by giving them thought in your own mind and consciousness. And that, I know, you would not have happen for all the world. Yet, you are figuratively putting your head in the lion's mouth when you even refer to them. You know frequency and attunement as it is explained to you in the mechanical processes of tuning the television and radio. Your mind functions exactly the same way. The dark forces in countless thousands of forms, both personal or entity formations, as well as the more abstract equivalents of thought-form bodies, are hovering on the fringe of your consciousness every moment of your life only waiting for a suitable opportunity to spring in like ravening wolves and beasts to tear you to pieces. Only because you have with you very strong protective forces have you been enabled to proceed along this line as far as you have and in the world and place in which you so presently occupy; and you are indeed occupying a very precarious position. And I am speaking of Ernest, too, and others who are seeking Truth and who have advanced to your somewhat equivalent position.

Remember frequency relationship, as it has been explained to you; you are walking with Ernest by virtue of all the things which you have done in past lives which would warrant such affiliations, just as He is so doing with you and others who are connected in a spiritual way with you. I can truly say that all of the great Masters who have ever lived upon the earth or who are known to mankind, and countless more, are working with you; give them full cooperation. You have a great responsibility, and the hell of Dante would be a kindergarten compared with that which you would pass through or into should you fail in your efforts at this time, and that would be a hell of your own making. I think I do not need to explain that point to you. You know the terrible repercussions of being able to see things in the Spiritual Worlds when you have passed from the flesh—neglected opportunities, ways in which you have unconsciously abused these privileges or opportunities, and various other different factors which are so relative to these things.

To one who is of the material world, those things are of little consequence at this moment; he is steeped in his sin and his iniquities; they are a part of him in his more formative parts of development. But after he has become spiritually quickened and passed beyond that dimension, he then chooses to return to the world, to work out not only his own past blocks but also that he could give his fellowman some advantage of the opportunity and of the knowledge and wisdom he has learned.

This is the first time that I have spoken through Ernest. This was intended to be something of a Scottish or Irish brogue; however, I see that I have reverted back into previous lifetimes where my dialect was even a little more Germanic accent. (Ruth expressed appreciation for his coming and he replied:) We have walked many pathways together in past lifetimes and as you know, there will be no one with whom you will form affiliations in this work or who will have part in this work who has not been preconditioned. Everyone has in the Spiritual Worlds chosen their part and they are working toward this common end.

You met one today, and she will be a big help to you both because she has many contacts. But these things are also cycular in movement, points passed in these various cycles of transpositions, so be joyful of this occasion, and make the most of all opportunities. You are doing wonderfully well and you have accomplished much in the face of tremendous odds; and the oppositions which you had to start with in this work were indeed tremendous, and these oppositions were from old lives you had lived in the past and their negative extractions.

Always think, if this will be helpful, that this is the true Ruth, the person who has obtained the idiom of many positive experiences of life in these countless thousands of lives she has lived, not only as a female but as a male, as a priest, a governor, a king, an empress and many other things, too, even a common servant girl or even a slave. Yet always, the positive extractions of these experiences which are optimistically slanted, do help to form and reciprocate in the polarity patterns with the higher self so that it, in turn, after an infinite number of lifetimes, becomes that great Higher Self. Always be conscious of it—that it is the wise person, the knowing person, the person who has retained the positive extractions of countless experiences and that that person is oscillating in the true infinite perspectives of infinite consciousness, and in that way, you will have no more false dispensations or desecrations against yourself.

P. Quimby

From The Master Jesus

To Ruth: Be happy and grateful that you have at last found a chosen spot and even happier that it has been with you to express in your own way a higher dedication, a fuller measure of faith and a greater realization of consciousness from within, for this Light which shines thus from within, shall be the Light which lights the world and the way in which every man, too shall find his pathway, his chosen spot. Fear not for the future, as ye have found a place where the roof of heaven has literally become one with you and that the very stars will light this home. These things, too, are all measures in their own way of that which you have expressed in your dedication and in your faithful way.

For one's path is indeed fraught with fears and apparitions of horror from the past and his flesh is rent with the thorns of great recriminations; and his night becomes a sleepless one—a chamber of horrors within himself if he has not served his fellow man and served the innermost desires, the longings and aspirations of the spiritual self, for thus only can man become a creative and participating element with the infinite consciousness of expression.

In common with those who are living about you in your new found place will be those, too, who are struggling or have in some way attained in their own way and in their own dimension, the necessary elements which will be necessary to also express what is to be given through you to the outside world. These and many more things shall be added for this is the way of the Infinite and it is constantly multiplying itself, even as the loaves and fishes; yea, and many more things both written and forever hidden.

Yes, blessed are they who serve and blessed are they who attain, yet in attainment find things which are not motivated from the desires of insecurity bred in the material world, but attainment of consciousness which breathes in and out within itself the very breath of life and all of the essences of creation weave its own shiny and lustrous garments to clothe its radiant form. Travel lightly, very lightly with the things of the earth, for in their shadows lurk the fangs of the viper, the renting fangs of the beasts which prey upon those who give way to their vicissitudes. Stay always in the radiant Light which comes from within and all things shall be added unto you.

(He gave no name. He never does, but I knew it could be none other than the Christ Consciousness Itself.)

The Wheel of Life

Just as water in its long journey
to return to its birthplace,
The sea turns the millstones
which grind the farmer's grist.

So does man's spirit in its longing
to return to its place of birth
Turns the wheel of life and its grist,
all of the things of each man's life.

The Everlasting Fount

Consider ye that all the fountains
of the earth and from all
The places hidden and unknown
from whence runneth the waters
which filleth the rivers
And their mightiness emptieth
into the great oceans
And of the clouds which filleth
the Heavens
Yet surely all of this is less
than one drop
To filleth the Place of Wisdom
in the Immortal Mind.

— Ernest L. Norman

A Discussion With Ruth

Ruth: Would you define please, problems caused by psychic shock, etc.?—Moderator: Tensions caused by shock from the outside, such as those caused by an operation, are one type. Psychic shocks are like static charges. They, in turn, reflect back some kind of galvanic action; they come back into the physical and cause pain from the psychic. In other words, the pain the person suffered before going to the hospital to have an operation reflects it back like an echo at this time when vibrations are set up where they are synchronized or harmonized through frequency relationship with the present, due to the state of mind the person is in or with other different wave forms being reflected through from the psychic.

That is a lesson in itself, isn't it? To better understand, let me say it is something like standing in a valley or on the other side of a mountain and shouting and hearing your echo. That is because the voice travels across, hits the side of the mountain and is reflected back to you. As sound travels so many feet per second, you get it delayed a second or two later, after it is reflected back to you. That's an echo. The frequency which we call pain, distress or various other emotional complexes of the mind, reflects back into the psyche and later on it is reflected into consciousness as pain. We are not concerned with time in this dimension; we are concerned with harmonic frequencies synchronized in frequency relationship, the same as we would have sound in pictures traveling at a certain rate of speed which, in this case, would be the element of frequency relationship. In the present, combined with various states of fear consciousness, fear assumes a new and different prospectus; similar to the message you were given on the concept of thought. Do you follow that now?

Ruth: Yes, and that's true; the added fears, the negation or fear one has adds to it?

Moderator: Yes, it colors it, just like adding vegetable coloring to a paint box to make it pink or green. So that is a new slant entirely, because there you are actually getting psychic pain that was caused, like an echo, from a previous circumstance. That reverts back to the more original concept which has been the trouble, such as cancer and various conditions in the psychic, existing as shocks of previous times; in fact, this can be related back hundreds or thousands of years from the past. In that case, they take on and assume more definite characteristics; cancer would be a definite characteristic rather than just pain itself. You could describe certain pain to a doctor, and at the same time, he could not vocalize it to you or tell you exactly what caused it because there would be no indication in the physical. There would be, however, with cancer. You might have some other organs which were defective or were not functioning correctly, which would be a direct reflection or oscillation from a psychic shock in a previous lifetime and which was very different than the present condition.

Ruth: Do you think that particular case was actually cancer?

Moderator: I still think that it was, because the person had it for so many years, however, I don't believe in cancer, period, because it is actually an external growth of psychic consciousness. Cancer is one thing that is even more predominantly psychic than any other condition a person could have, because it is like a toadstool growing on a stump, you might say. It is a physical growth of a psychic condition. You could put it that way if you want, but going right down into the actual cause, it merely means that the various wave forms portray a certain intelligence quotient, and are the motivating force of every atom, and which sums up the total, basic, motivating force of the cell! This so-called cancer is therefore unintelligent. It has no relationship to its present consciousness in the body as a function. It could be intelligent in any other direction because intelligence is relative, but it is not intelligent in relationship to the body any longer.

You can go right back into that abstraction and say, "Well, what is sin?" Who can tell what is sin? We can only compare sin by what we think is good. We could say, in symbology, if we were psychically standing on our head, then sin would be good, and good would be evil; so what basis or what platform do we use to evaluate these things? The only predetermining element there in the whole situation is what we call the element of progression or constructivism. We can say that we can establish any particular chain of circumstances or events, one leading to another, and each succeeding event rebuilds, recapitalizes and enlarges the perspective or horizons of consciousness in past experience. The past experience under those circumstances can be called good, but by the same token, if it reverts consciousness, then it is supposed to be evil.

There we are getting into that abstraction which Einstein was formulating in the fourth dimension or the venturi tube. Consciousness can expand in two directions as far as our present position is concerned. We can expand outwardly and downwardly. We can see infinitely into the subastral regions in what we call sin or error, or we can expand the other way, but that is getting beyond the realm where human mind can understand. We don't have the proper mechanism; we don't have the proper hookup or function into higher planes of consciousness except on very rare occasions in such instances when we have what we call visions. When we see visions, they become so incomprehensible to us, that as far as our conscious mind is concerned they mean absolutely nothing, but to our infinite consciousness they can mean everything. That relates, of course, to future events and circumstances which we are yet to experience from the conscious mind. After the vision, the conscious mind has been focused, we'll say, in an entirely different, a new or better direction, assuming of course, the vision was inspired or came from the higher sources of the mind or sources of the psychic body—the psychic mind which we call the Superconsciousness.

These are rather deep abstractions, but we can go along with these ideas and these thoughts so that you can formulate some very definite conclusions for your own self; all steps in the right direction, because it all resolves itself into understanding what Infinity consists of. At the present moment, we have a third-dimensional mind; that is, a mind that is more or less circumscribed because it is functional to the extent that it forms one end of the pole—a polarity of consciousness. The other end of the polarity of that consciousness is, of course, any one or a number of different levels of intelligence which are energizing forces of wave forms in the psychic. The further we go into the higher realms of consciousness in the psychic, the greater the distance becomes in its relationship to the external mind—the so-called human mind—in the balance of what we call polarities, because of the differences in the rates of vibrations or frequencies to which the physical mind is capable of responding.

In other words, we couldn't expect our television set to respond to frequencies for which it was not designed. If we get into frequencies beyond either one side or the other of the thirteen channels, that set will not respond. The radio responds to a different set of frequencies than does the television, and so forth, or the conscious, human mind is only capable of responding to a certain limited perspective of frequencies in relationship to the polarity to the psychic. However, there is an important thing to remember: all that can be changed through evolution by what you call spiritual evolution or mental evolution! Regarding the clairvoyant, or a person who is able to attune to higher thought forms or strata of consciousness in the psychic self, this simply means that the person had passed through a longer period of evolution of training or consciousness into the higher realms of consciousness. He has developed his mind, just as an athlete develops his body, so his mind, therefore, functions differently than a person who has a mind which is a polarity with the lower strata of consciousness of the psychic, which is commonly referred to as the subconscious.

Present, modern-day psychiatry is only concerned with one level of intelligence as far as the psychic body is concerned, which they refer to as the subconscious. The psychiatrist can't even remotely envision that a human being has any more different relationships or strata of intelligence or states of consciousness as far as the psyche is concerned, no more so than he can picture any other particular thing of a spiritual nature. He is only concerned with the reactionary contents such as might be portrayed with a certain crude set of symbolisms which he calls reflexes—reflexes either hypnotically, autosuggestively, or even consciously displayed—which are elemental, as they only relate to the subconscious. With one exception is this true—and that is the effort and research the Rhines at Duke University have made—that man has never gotten beyond the realm of the subconscious, as far as the psyche is concerned. They have arrived into what you call strata or dimensions of consciousness of the psyche which were of the mental caliber, we might say, the mental body or mental mind of the psyche where mental telepathy, telekinesis, psychokinesis, etc., could be definitely proved beyond the element of doubt or chance, or any other particular thing which the scientist might call pseudo.

Of course, they are always bringing up the thing, that dice could be made to fall a certain way, which is psychokinesis—a power of the mind over matter—and very crude in its basic concept because, in the first place, no one needs to voluntarily express a thought in any direction if he is properly attuned to the higher strata of consciousness of the psychic self. If he is, then the conscious mind is always reflecting a power which has dominion over everything. That was so ably expressed by Jesus. There was no particular situation which Jesus came upon which could not be handled very adequately by the power of His mind, simply because the mind functioned on an entirely different level of consciousness with the psyche than those about him.

That is the secret which distinguishes great people from the average earth man; people who have power or dominion over other people simply because their mind functions in a higher level of consciousness or in a different level of consciousness yet not necessarily higher, because there again we don't want to misconstrue what we call good or evil, or higher or lower. Some of the greatest and most powerful men in history were, by our common standards, very evil people, such as Hitler or any one of the hundreds that you can think of, but they exercised great power and dominion over other people. Many people have wondered why this is so. Well, you might say they had whole hosts or legions of astral forces, evil ones, working for them. That was true, too, to a certain degree, the way in which the person's mind functioned as a telekinetic or psychokinetic force in the regulation or the extension of force, power or dominion over the minds of other people.

That was the appeal, we will say, or the telekinetic force or appeal which Gandhi expressed through tele-

96

portation, telepathy or whatever particular channel you want to call it, into the minds of hundreds of millions of people who lived in India. We find that appeal was through the more spiritual side of their nature where, in the case of Hitler, it inspired the lower or subastral parts of the subconscious; it inspired all the evil to the last degree. In other words, that became a lustful escape mechanism, and in the case of Gandhi, it became a religious escape mechanism, because wherever we find man, even before he assumed some of the more primitive, elemental forms of life as we find it on the earth today, he has always been possessed with that one dominating fear—the fear of death.

The fear of death is the primary fear which has dominated man and still does, even though he finds himself a priest or someone who is supposed to know eternal life; yet the fear of death dominates the subconscious to such an extent that it is the primary evil of all fears. It could be called the Satan of all fears. Learning to rationalize these things is important because the old fear of death is inspired from out the various elemental fears which raised the storms around man in his first of the many countless thousands of reincarnations, whether it was the roar of the mastodon, the saber-toothed tiger or the roar of the thunderstorm, the bolts of lightning, the roar of the river, the rumble of the volcano. Whatever that fear was, it spelled one thing to him—the threat, the danger of extinction. He had nothing at all to support any belief that he could survive beyond that time, simply because this was one of the first ways in which he began to know life; he had to learn it and begin it in primitive surroundings because he was primarily a primitive creature. He was still two-thirds animal and very little man. He couldn't have begun that evolution in a high state of consciousness because it would be like taking a full-grown gorilla into the living room of your own home.

He just wouldn't fit. So he had to begin that evolution in the environment to which he had grown accustomed, just as we, in a physical sense, are accustomed to breathing oxygen and nitrogen. When we introduce helium into nitrogen and the mixture of oxygen, then our vocal chords are affected, so we see the result of environment, even in the chemical contents of the air, as far as molecular structures are concerned.

This reverts back to our original and basic concept, that man, in whatever particular plane or strata of consciousness we find him—and that goes for whether he is a primitive aborigine in the jungle or whether he is a highly developed scientist, priest or any other person you find who, you might say, expresses the highest form of mentality on the earth today—he is still basically an elemental creature, an element of creation, simply because he is expressing a small part of one particular moment, a certain definite polarity in a negative tense with the positive polarity of consciousness which we call the Superconsciousness. That positive polarity is the sum and total of all Infinity and all of what Infinity is, or that which most people commonly suppose God, for they see God as a personalized being, but tremendously over-sized as far as they might envision, beyond the limits of prospectus of their own horizons of consciousness. They say God is omnipotent, omnipresent or they can use any other particular superlatives to describe God, but the average person is still a pagan because he is deifying God in an external dimension. He has rationalized only very faintly or very crudely that God is the sum and total of everything. He is not a personal being. That doesn't mean as much to the average person, the man on the street, as it does to you and me. God is still the white-robed Santa Claus sitting on that seat up there is the sky somewhere, with the Book of Life, the way the average person, especially a Christian, is liable to rationalize God. I believe the people in India, as far as they are able to read or able to get information, have a much clearer idea or concept of what God is. Other people in the world do more so, too, than we do as Christians; in fact, the Indians or the aborigines in the jungle sometimes have a much closer approach to God than the so-called Christian because, at least in part, they see God in everything. The great Manitou, the Spirit in the sky, the earth, the air, everything was God to the Indian. The great Manitou was the Spirit that energized everything.

So it all reverts back again to our original concept of understanding energy and our relationship to what we call mass, what the sustaining, motivating forces are, how they are concerned with polarities, because consciousness always maintains consciousness, simply because it is two polarities; they are vibrating in harmony with each other as two opposite and extreme perspectives. Call them good and evil, if you wish; call them material and spiritual. Whatever you call them, you will find two opposite polarities in all forms of consciousness because of the fact that these are dynamic movements of consciousness in wave forms or cycular motions in every dimension with which they are affiliated. Again you are going to find they are linked and relinked in a never-ending chain and succession of chains in cycular patterns of consciousness to everything in the Infinite Universe. In that way, man becomes the sum and total of the Infinite Mind of God.

Whether we contain the idea of the concept of Superconsciousness into the elemental form of a cycle which contains all of Infinity, or whether we are relinked as dynamic, moving polarities of consciousness to every other polarity of consciousness in the infinite cosmogony makes little or no difference; the abstraction is the same. We are the sum and total of all Intelligence, as far as that is concerned, because we have the internal and external apparatus of what we call thinking or consciousness to do that particular thing for us or with us consciously. Any other way, it is a subconscious virtue or attribute in which we function in a normal process of life which is commonly referred to as reactionary, and which is the lot of at least ninety-nine and nine-tenths percent of all of the people on the face of the earth. That remains so until we become spiritually quickened and conscious of the fact that the priesthood of the temple or somebody else does not do these things for us but we do them for ourselves. Therefore, we become conscious of these great powers which are inherently ours, simply because we are opposite or activating polarities or actual, participating polarities of consciousness into the sum and total of the Infinite Cosmogony.

That is the great secret of life; that is the struggle through all of the carnal worlds of consciousness with every individual, to arrive at that point where he realizes the full impact of that truth. Only when man arrives at that point does he find, like Buddha says, his Nirvana or, as Jesus said, we have found the Kingdom of Heaven Within; we have found the Father which doeth all things. Even more important than that, we know what God is; we know in what way we are a part of that God. This is only understood, evaluated and can only be envisioned when a person understands what energy is, what principle energy moves in, what dimensions of consciousness of perspectives, the various frequency relationships, harmonic structures, cycular patterns and various other different factors which are concerned. Only then can man conceive his son-ship with God. Above all else, we must always be conscious of polarities; that, in itself, is a most important part to remember in sustaining our consciousness into whatever particular thing we are trying to think about or envision.

This is all part of the psychology of the future, the

Aquarian Age, of which man will become conscious. I doubt very much whether it will ever be in this world. I think, as far as the future of the Aquarian age is concerned, that, in itself, will relate any individual who becomes conscious of spiritual progression and passes that point of which we have just spoken—inward consciousness—which is the dynamic, participating element of polarity into Infinite Consciousness—then he has reached that point where he actually enters into the Golden Age or the Spiritual Age. When he does that, he will, automatically, without fuss and without fanfare or drum rolls, pass from these earthly planes of consciousness. He will pass from the material into the spiritual and return no more.

That is primarily what is meant in the Bible; that all the prophecies or perhaps any references to the Golden Age in any of these prophecies relate or refer, shall we say, to a certain group—unnamed and un-numbered perhaps, except in Spiritual Consciousness—people who have been traveling through a vast number of lifetimes and an infinite number of dimensions of consciousness until they have arrived at that point of consciousness—just as you and I are arriving at it in this lifetime and in this day, which will make it almost impossible for us to return to this form of life again in the future. So we will be citizens in the new Aquarian Age! We aren't particularly concerned where that is going to be or when it is going to be, because we know where it takes place will be, by far, a higher spiritual plane of consciousness and a far greater world than is this one. As far as 'when' is concerned, we are not con-cerned with that either, because as our basic concept relates to Infinity, time has no part or place in any of these future dimensions of consciousness, except in this one particular one in which we are engrossed at present. In the future, these things will be in cycular patterns which will relate us to the accomplishment of consciousness in some particular dimension or strata, or any other particular factor you wish to call it.

The concepts we are talking about in this time and place could be said in different ways or in different words; for instance, like Nietzsche, who talked of the Superman. The Superman we have talked about is the sum and total of consciousness after man has passed, as a polarity, through a great number of dimensions of consciousness, and after he has become aware of these basic frequency structures which relate him to Infinity. He has become aware of this because, again, he has maintained polarity in two different dimensions; first, through the experience dimension where he retained the idiom of experience in all its relationships. These, in turn, as harmonic structures, related him to an equal number of dimensions of consciousness in another form and another polarity in a different place. And so the pattern runs. You see the complexity these concepts can begin to assume.

Only the people who have gone into mathematics and electronics and have worked with such things as sine wave generators, oscilloscopes and other instruments can barely begin to comprehend the complexity of what we are trying to evaluate. Regarding our function of life in the common language in which we could interpret it as wave forms and harmonic structures, frequency relationships, spectrums, dimensions, etc., we are, in such concepts, really beginning to conceive life as it should be realized. So, it is a wonderful science and really a challenge to anyone who wants to throw aside all of this balderdash and chicanery that is running riot and rampant through the various channels of advertising media which we see about us every day. They use innuendos, pseudonyms and various other things that hint at this and name fictional names, which are sometimes composite forms of all idiomatic expressions, concepts which have been used in occultisms for thousands of years, merely to impress people with a false sense of mysticism.

People, as a whole, like to worship; they like to engage in idolatry. Even when Moses turned his back on the Israelites for just a few days, they set up the golden calf. And why do they like to do these things? Simply because they are escape mechanisms. Morally, people are shiftless and lazy. Spiritually, they are even worse. They are only forced into doing these things because they have a great and vast system of laws, regulations and various social structures, moral incriminations, laws in which the average person is literally squeezed and pushed into a certain channel as a function of some socialized system, whether we call it a democracy or capitalism, communism or whatever particular thing we want to call it. The conditions and their causations and effects are exactly similar. Because in the first place, we have a population that is dissimilar in exact proportion to the number of people of which it is composed. No two people are the same, but the concept of government, social structures, etc., as we know them on the earth today, relegate people to one size or hole in the screen of life. Everybody is made the same. If one person becomes greater than anyone else, has something greater, sees or does something greater, he is immediately ostracized. Countless thousands of people rise up against him in an attempt to pull him down. This is directly in contrast and presents a very strange paradox to a psychological escape mechanism which he calls idolatry. He likes to worship but he loves to tear his heroes down. Isn't that strange?

Once you start to equate these things, as we say, in the reactionary sense from such psychological values that are contained in the reactionaryism—which reverts back to the subconscious and the various primal motivating forces which were compounded into that subconscious matrix—man begins to lift his consciousness. This is so because man is not a thinking machine, he does not possess a thinking mind. If he did, he wouldn't do all these strange and paradoxical things that he does. A man or woman will love each other until they can possess each other, then love becomes something which is like a cage to confine an animal and treat it as such. The same thing goes for many other different aspects of life. Men will idolize a national hero like Eisenhower or any other President before they elect him, and before he is in office a year, he will have ten million bitter enemies, the bitterest of his enemies sometimes those who helped elect him!

It was the same 2000 years ago. Jesus rode into the gates of Jerusalem on the donkey. They waved palm branches and hailed him as a Messiah, and a few days later, the same people helped to crucify him. So they really never begin to get anywhere; they never stop being reactionary, like creatures of the jungle. He is influenced as is the leaf on the tree by the wind. He is constantly influenced by the will of countless other people who are in that same dimension of consciousness. Man becomes subjective to every particular force or pressure which is concurrently stirring in the world about him! He really isn't a human being in the sense he believes himself to be, for he is just an atom, just a piece of driftwood, something which has a certain mechanical function, to work, to pay taxes and to be one more of the digits on the census system. But as a thinker, man never really becomes a thinker until he to the various reactionary elemental forces which are swirling in this world of consciousness about him. The only way man can refuse to so live is when he learns that there is something better, when he learns what he really is, or what he can become and what he can have; when he learns that he is an actual polarity, in every sense of the word, with the whole Infinite Cosmogony, then he will refuse to react, either as a whipped cur or some great general, a political leader, or whatever other particular expression of life you might expect any human being to assume.

Up until all of that, with the fanfare and falderal, man now begins to assume a position in that consciousness as a polarity in which he is oscillating with the Infinite. One man put it thus—and he probably didn't even consciously know what he was really talking about—"In tune with the Infinite." When man actually becomes "in tune" with that Infinite, he will pass from that reactionary plane of life. Oh, it may take him a few years, maybe even a few lifetimes or a few hundred lifetimes. It depends upon how much of that attunement man can achieve at one time or in one lifetime, and just how much moral virtue, stamina and spiritual integrity he may express; how much of the courage of his convictions he can assume in his own personal relationship to the reactionary forces he is leaving behind.

In the great Spiritual Worlds which are ahead are people who begin to think or who become consciously attuned with the Infinite, those people are vastly different than the people we find in the material world today. They are different in so many ways, it is very difficult to describe what they are. They aren't the people who go into seance rooms or hover close around in some of these astral planes of consciousness, close to the surface of the earth with great psychotic pressures, because when people die, they don't lose these various psychological malformations and aberrations; they become only increasingly conscious of them. And those people who hover around close to the surface of the earth, slip in and out of the physical plane wherever they can. They influence, they badger, they beg, they lie, they cheat, they steal, in every psychic form of consciousness they can possibly imagine. But we are not speaking of those people of the lower astral because they aren't any more advanced than the rest of the people on earth or those whom they are coercing in that particular way. We're talking about those people who have gone on—those who have progressed in their evolution.

I am sure we have tried to give some of that information in the books; we are giving it to the people in a way in which they can understand it. I feel equally sure, when the day and the time comes when we, ourselves as individuals go into those worlds, and, shall I say, become as physically conscious of the Higher World as we are of this one at this present time, these experiences and expressions are all going to be very different. I know that; I am very consciously aware of it, and that is the reason I always have the feeling the books are inadequate. They fall far short of the mark, and I feel sure that feeling is shared universally by the ones who participated in bringing these books into consciousness. But it is the only way in which these "little children" can begin to learn. It is justified, at least on this premise, that if they can only begin to become consciously aware of something beyond this treadmill they're on, then there is hope for them. Presently, they certainly have had no hope, no realization; they have been fed this fodder and this garbage which has assumed all kinds of forms and shapes, always headed by one name, "spiritual truth." It is anything but that. Every one has become a blind alley, a place littered with garbage cans and refuse, from heaven only knows how many past ages—a way in which they have to reach the end. Then it ceases to be of any more value and they have to turn their faces around and tread their way back out and 'unlearn it' all over again. Each time they have to do that, they become more sick, they become more malformed, they become more dissolute, less of that moral integrity, more of the physical weaknesses and infirmities and disease which always accompanies these things.

Yet there is always one thing, one difference as to just how far people can get beyond this reactionaryism which every person carries in their mind (which they call their philosophy of life)—this rubbish pile, this pail full of garbage, because that's all it really is. It's bits and litter and debris, this, that and the other they pick up from the moment they left the cradle—not only from this life but heaven knows how many other former lives! And they've accepted it because it had one or two elements which, when anything possesses those elements, they would accept it. Even if it was a dragon that breathed fire, if that dragon possessed the same elements, they would accept that dragon into their own home and it would burn them down just as quickly. And those two elements are mysticism and idolatry! (Mysticism that appealed to them from back in the old —old days when they were frightened by the elemental forces of which they knew nothing.) That became the first, the prime, motivating fear of their life—fear of the mystical, fear of the unknown.

Call it wave forms of intelligence, call it cycular patterns, call it whatever you wish, but it's still there. It has built a fence around so many people—a fence man is quite powerless and incapable of climbing over and destroying, or even looking through; a fence built of stones and intangibles. Even Jesus would have a hard time tearing some of those fences down; impossible, because even though the individual knows these things are hemming him in, he would become very angry at having them removed because he has nothing else to support the loss of something familiar, nothing else with which to rationalize those things that support the structures of his life, evil as they are! Yes, this has been the way of the earthman for lo, countless thousands of years. However, now he does have the science of Unarius to teach him the true way to learn to live a more progressive evolution—and eternally.

A concept of creation of man is when we find an aggregation or a collective aggregation of intelligence which can be identified as an ego which can survive in the ego form as a psychic body from one life to another. We have a human being because the individual lives and relives in that psychic consciousness which we identify as the ego, from the culminative expressions of previous lives which have formed that nucleus or that cell of consciousness which lives and relives. That, in itself, proves reincarnation. Anything which we call an instinct, whether it is two (so-called) instincts which were given by Freud, or any number which have been included in various other different psychological concepts, it makes no difference because, even if it is one or a thousand instincts, it is the direct admission that psychology, in that expression, admits reincarnation! It has to, because every one of those instincts is a dimension of consciousness which we call fear, which has survived in that idiom of the ego through countless thousands of lives since the first beginning of that individual when he began to retain the idiom of the ego from one life to the next.

And that's what distinguishes a man from a beast or a bird, in that sense, because it is through this matrix of wave forms which form the nucleus or the cell of the psychic body, we call it, in the expressions of dynamic energy reflecting into various polarities which link and relink him with the whole Infinite Cosmogony. It is even more important to remember that ego survives from one life to the next and that his various different expressions of temperament, intelligence, etc., cannot and have never been justified in a genetical sense; they can't be, but they can be fully justified and can be fully explained when a person understands reincarnation. Because, for one thing, the very fear of death itself which concerns the greatest portion of the population of the earth is an old primary fear which man first arrived at in the beginning of his evolution, in what we might call the human being. That one fear alone—the fear of survival—is sufficient to prove reincarnation. Any instinct, large or small, call it the fear of survival, the fear of death, call it sex, whatever you wish, the primary, motivating fears or concepts are things which were first brought into consciousness in the most ancient of any one man's lives.

Sex, itself, is the procreative instinct which, by the same token, could relate man back even to such a substance as a fissionable amoeba, but sex, in itself, does not retain the idiom of consciousness in the ego sense. The personality, as it is retained as one human being who goes from one life to another, still retains the elements of the personality characteristics which are constantly being developed and redeveloped in the psychic body through experience.

Now that is a very good explanation on that one particular concept, a little point I have gone around with for many years and it is very clear to me now. We have to draw the line in what we call ego consciousness as it contains itself into any one particular individual. Whether he is John Smith, Joe Doakes, or whoever he is, we carry that particular person's ego on back through countless lives to that time when these began, through the process as a collective amalgamation of reactionary fears which survived from one life to the next as an ego, as a consciousness, as it is personified with the individual. Looking at it as it factually exists and looking at it also as a negative polarity with a positive consciousness in another relationship, it begins this collective process where these things accumulate together to form the conglomerate psychic body or consciousness, or which is the end of the polarity of Mind Consciousness, or Infinite Consciousness. In that particular negative experience, the psychic body lives and relives from one dimension and from one lifetime to another.

Now, going on up until we reach that point where he begins making a conjunction with the Infinite, or what we have spoken of a while ago when this person—whether he is Joe Doakes or John Smith—begins to be conscious of himself in another way—not as just a taxpayer or cog in a wheel of some industrial empire or a soldier in some Caesar's army, or some other particular way of life, he becomes conscious of the fact there is an Infinity around him; there are an unlimited number of dimensions of consciousness. When he becomes conscious of that fact, then he begins to gradually evolve into that state of being in tune with the Infinite (in any one or a hundred or a thousand lifetimes, however long it takes him). That is the time when he actually gets past the material world; he is no longer a part of it. He then becomes the negative part of this polarity consciousness which we call the psychic body or subconscious. Then, in turn, he begins to change form and assumes different relationships with these higher strata or planes or dimensions. The old, baser, materialistic waveforms are gradually replaced, the psyche rebuilt into a spiritual (if you want to call it such) aggregate or conglomeration of a vast and infinite number of forms of consciousness as polarities in themselves with the whole Infinite Cosmogony. That way he is a spiritual being, no longer a fleshly, materialistic man living in some earth plane or reincarnating in some comparatively low spiritual plane of relationship. Now, he has become one of those people we talked about, as Avatars or Masters, Initiates, Logoi, Archangels, etc., depending on how far he has progressed into that relationship where the constant and succeeding changes in the psychic self or the spiritual body are being replaced with higher and more infinite numbers of wave forms or aggregates of consciousness. This, in turn, relates him to ever-increasingly higher numbers of planes of consciousness, ad infinitum.

So you see, these small minds and the patter of these small feet encourage me sometimes to get out articles like we have today. They go into these various different garbage cans, receptacles that people have used to express their concepts of life, in many times and places. They get this collective mass of potpourri together and they call it a proof, either for or against reincarnation. They haven't proved anything at all. They haven't postulated a single intelligent concept in the whole verbiage of mental garbage. The most peculiar part of the whole thing is why. Why don't we have a channel of expression today? I am sure there must be people who can express something a little better than that. Aren't they permitted, or aren't channels open that they can express something? Sometimes they reflect almost the ravings of a psychotic maniac. This destructive emphasis on everything they do is degenerative; there is nothing constructive in all they express by these vituperous insults—nothing at all constructive. They get nowhere in these primary and motivating laws of metaphysics.

The greatest of all concepts a person can arrive at in any particular place in his evolution, no matter how spiritually conscious he becomes, particularly so in the material terms of life, is to realize the importance and the fact that when his mind begins to function in higher spiritual levels, all of these so-called methods and various other elements of metaphysics these people have bandied around through the years become absolutely child like and useless, like little girls playing with mud pies and boys with tin soldiers. When our mind begins to function in a higher spiritual plane of consciousness where all things begin to assume that infinite relationship with the Infinite, that, in itself, begins the levitating, if you want to call it that, or projecting through the process of psychokinetics; not only into any conscious objectivism which the individual might portray in his immediate perspective, but into an infinite number of relationships in direct ratio and proportion of intelligence which he has portrayed into everything which is not of the same intelligent level— the lower intelligent level. It has a leavening action, you might say, in the evil or destructive.

For instance, Ruth, we might say this talk that took place between you and me will be listened to by other people on this tape possibly in the future. So what has been going on in these immediate moments between you and me in the past hour or so has been of benefit to everything in the infinite universe—if it is on a lower level of intelligence than what we have been expressing! Now, it has related itself through frequency relationship, in the line of harmonic structures portraying as it does a certain constructive intelligence or force, if you want to use that word. It has a leveling action, a counteraction to any evil thing or anything which we might call in a lower state or plane of intelligence in the whole infinite universe or cosmogony. That's a very important concept to realize. The sum and total of all things so expressed is, to a large measure, the only thing that has levitated man from the caveman days up to the present moment. It is one of the greatest of all single forces or elements of force, we might term it, or power which has ever begun to elevate man into his present position—that is the sharing of the good and common with the unseen. Even though the whole universe or the cosmogony is unseen, unheard and unfelt, in the moment in which good can be shared or given a common level is the ultimate in man's ability to express himself infinitely when he is so attuned.

And so, may all efforts in attunement of future Unarius readers be of the most high states.

Philosophy by Lahamahasa

To Ruth: You were speaking of personal identities and that there were many thoughts which have been passing through your mind and especially leading up to the particular point; when one becomes wise in the ways of the Infinite. If one uses identity as a personal entity in relationships which would inflate the ego or would further stimulate the person to greater efforts in directions which could be called such ways and manners and means of acquiring personal gains (and when I speak of personal gains, I mean such things as are most valuable to the egocentric and, in particular, to his sense of well-being and importance to himself), he becomes a small planet. In his own orbit he is revolving around the seat of the throne of all intelligence, and so, it seems to him that he is only waiting a suitable opportunity when he, in his superegocentric consciousness can rush in, at the most opportune moment, and occupy this seat of Infinite Wisdom.

Unfortunately for him at least and fortunately for the rest of humanity, such a condition never happens. For, in the sense of infinite proportions, those who are most concerned with their own position and in a way and manner and means which can be equated from such superficial or reactionary values from the subconscious which are stimulated from past lifetimes—are living in bestial forms of consciousness which are much akin to those of the beasts of the fields and forests upon which he preys for his food. Thus, identities become relevant and valuable to him. His name is inscribed in his holy of holies; he worships at the altar of his own false superstructures which he considers the most valuable thing in his possession, namely, how much better he is than his neighbor. He must drive a

113

better car; he must live in a better house; he must do so-and-so so much better; yet he knows, just as he is doing these things, that they are only relative and for one person such as himself, who is thus doing these things and concerning himself with the attributes of the material life, that there are many thousands more who are in similar conditions perhaps better—some perhaps less so. But all are reacting from these same constituents of reactionary values which have been so firmly posed in that urge and that struggle for survival from the more primitive days of his existence, and one of the last surviving bastions, shall I say, of these various karmic structures which are necessary for his existence on the physical plane is his own personal identity.

However, please do not misunderstand what I am about to say. We are separating in the common tempo of living as it is expressed by the materialist and in the centralization or self-focusing of all objectivism to that point in evolution where such objectivism and focusing and such relationships assume infinite proportions; where he can realize that all acts of consciousness, which he performs within the dominion of his own perspectives are, in themselves, multiplied seven times seven as useful and added constituents of intelligent energy formations which, shall we say, like rays from the sun, permeate and warm the lower regions which are lacking in these necessities of life. While the sun seems to give the life-infusing rays which enable plant and animal life to function upon the earth yet, the earth itself and all forms of life could easily cease to exist and in a moment's notice were it not for the great influxes of intelligence which stem into all of these forms of life from the higher dimensions. All of these things do not say that "I am John Jones," or that "I am Bob Smith." They are concerned only that they are elements of Infinite Consciousness and, as such, are most

114

content and happy to live in the warm rays of that Intelligence.

So it matters not, little Sister, that you are traveling in the company of One who perhaps, and I could very easily affirm what you have established in your mind as you have seen Him traveling in the Higher dimensions as Zoroaster or Zarathustra, as you more properly know Him—the scientist from Lemuria who has created the living flames for those who are now known as the Parsees, and only 124,000 of them remain in the earth-plane consciousness and, that you met one of these in a brother priest in the Temple of Flame about two years ago, did you not?

You see, Lahamahasa knows many things. Who is Lahamahasa, you say? He is one who has had the privilege of keeping company of One, just as you are so keeping company with Him at this present time. You are performing a very valuable and functional purpose in the world with Him; and may I say to you that the last thing which would be conceivable in the mind of Him whom you call husband is that He would wish to demonstrate supernormal phenomena power to anyone. And, although you know He could do these things quite easily (and He does them at all times with His students and to those who come in contact with Him) yet, even the very nature of their intelligence is changed also, so they do not think that it is miraculous; but, as you suggested too, it is an activation into consciousness, of that which they know in the higher self.

Yes, little Sister, I can assure you that even those who are called Masters and who are worshipped and venerated by more than nine hundred million people who call themselves Christians, that these people in themselves, have been taught and taught by the Lemurian Masters. For it is the planet Lemuria itself which is one of the great Centers of learning for hundreds of millions of terrestrial planets in the great universe and that many of those who journey through the starry pathways in the sky in the flying saucers, as they are called, are being taught by the Lemurians into the higher sciences of life. Yet rest assured that in the common understanding of energy and that as the world in itself has its own basic frequency—if I can thus speak so—it would, in effect, be quite impossible for any regeneration of such harmonic structures which would be conducive to a higher way of life than that which you have or are presently seeing life so expressed on the earth, that the billions of people who swarm the little planet are merely in the very lowest and elemental stages of manhood. They are no more—or no less—than they have been for hundreds of thousands and even millions of years.

And the earth shall remain thus and so, passing through cycles which to some degree create a rise or fall of certain civilizations or epochs of time wherein certain differentiations of intelligence are manifest. Thus it was in the days of the Romans, the Greeks, the Carthaginians, the more primitive Hindu and Assyrian civilizations, even the Inca and the Mayan and those whom the Lemurians left behind in the Atlantean epoch. And, each time they have come and each time they have gone, they have left behind them the pure dispensation of life; yet, you can easily see how the more primitive and savage forms of mankind which surround you, have succeeded in destroying these things.

Even today, your leaders in the White House, as well as the most common citizens in your great country, are shaking in their boots at the prospect that war is imminent because of a crisis which has arisen over a few hundred square miles of territory. Simply because a few people refuse to let their egos be deflated, they must make hundreds of millions of people suffer great physical distress and a possibility of physical destruction and termination of the present of their reincarna-

116

tive pattern; yes, even possibly rendering your planet untenable for mankind for a hundred or more years until it cools off simply because they are not grown up. They are still savages traveling through the jungle trails.

You call them "President", you call them "Congressmen", you call them other leaders but they are still savages. They have better ways of butchering their fellowman, but they are still savages. You are indeed fortunate that you are traveling in a small sphere or globe of energy through this material world which shields you to a great extent from the austerities which most people suffer in that karmic world which is about you. You are most fortunate to be bathed in that Radiant Effulgence of Light which comes from the Higher Planes at all times and to have at your disposal and within your hands the key to Heaven and hell if I can put it so. What other person on the earth holds such a position? No, there is nothing in the mind of Him who serves the Infinite and who walks by your side in that terrestrial plane that He cannot contact, that He cannot do, if He so desired it so.

Yes, it is a problem which has come to many who have visited this little planet, and even as the One called Jesus hung on the cross, He was sorely tempted to put aside His material disguise and to call upon the powers of the Higher Worlds to deliver Him. Yet He knew even then that in His great plan, in order for anyone at all to believe in Him for a period of two or more thousand years, that they could only do so under a blood-smeared banner of hate and lust, so low is the consciousness of those who call themselves the priests and who are supposed to be dispensing His brotherly love. And yet, we will not have it so at this time, nor will He have it so who is there with you, and we respect Him in His wishes, and admire Him in His fortitude for donning the mantle of flesh and enduring the vicissitudes of the material world. Yet, it is there that a supremacy of wisdom and knowledge must assert itself as it is doing and proving itself to the fortunate few who have prepared themselves for His coming.

He has no wish nor desire to tell the world that He is so-and-so, except that He realizes that there may be a few who must know His name, as it is called and believe for the sake and the knowing that they have seen Him and talked to Him in other worlds and in other different times. And when they do so, the Light is let into the windows of their lives and they are changed as you have seen them changed—and as you have been changed. For these are the men who have taught others on your world, the things which they have left behind them and for which men worship and venerate them to this very day. Yet the greatness of this is contained in the reticent way in which they have merged themselves into the Infinite Consciousness and that no man knows who they are to this day save possibly you and your husband. Nor, if they do so, they have an inkling.

Again, I can only repeat what others have said before me. Rest assured, as I know that you shall do, that whatever He came to do will be completed at this time and in this cycle. For if you know Him as I know Him, it would be to know many others who are spoken of in hushed whispers. Yet, look not upon Him and the frailties of the flesh in which He must express Himself through the world, for that is the way He has chosen, and that is the way it must be, simply because of reasons best known to Himself in the variances and in the valences of certain psychological values (if you can call them such) which enable Him to maintain a vehicular form of flesh on your earth. And yet the cycle must swing both ways and so He finds Himself today, perhaps in the most difficult of all positions; to maintain something in a material world which is entirely foreign to His nature, which He does not believe in, yet He must fully understand and work with. How else could it be? No, I shall not say that He does not believe in it for there are those who may read these words later who would not understand what I mean. Do you know? (Yes, I do.) He does not believe in them simply because they are contrary to a higher form of life; but He understands thoroughly that in the development of an embryo spiritual person, these are all additives; they are condiments and they are constituents which, like the foundation and the building blocks in a building, must be laid before the tower and the belfry wherein the bells are hung. But even that and the words which I must use are but crude facsimiles perhaps, of what I am trying to convey to you. Yes, indeed, dear Sister, yours is a most important position, but treat it always with the greatest respect and dedication of purpose, and, as I can only stress what you well know by now, perhaps these words are superficial. But I can only stress their importance; that He must be shielded as far as possible and especially, must He be given love and understanding.

To you, because you now have ascended and you are taking your position in the creative scheme and movement of things which are progressing forward and you are ceasing to be the reactionary individual who was working out karma a few short years ago, indeed, be very careful when you feel any intemperance coming to you, that you will know, that is what our friend, Jacobi (Einstein) spoke to you about many times— perhaps a bit forcibly.

Perhaps Lahamahasa may or may not be able to visit with you in the future. I hope that I shall be able to continue a series of narratives in the more philosophical vein of teaching which would perhaps be understandable even more so to those who were sufficiently indoctrinated into the higher concepts of energy formations. Such things must always be made necessary to those and these things must always be continually added. You will not, in your span of lifetime, see—nor has the world seen, since the Lemurians—a way of interpretation of life such as is expressed from these Higher planets; because, in Lemuria itself, one finds a way and an outlet for what we know as the dynamic and the motivating, the moving forces of the Infinite Creation as it concerns mankind individually and collectively.

And, I have only one little observation left which I may be permitted to make at this time; that I am wondering, within myself, just what is conceived to be the terminating point—how far the physical body can be projected into the Spiritual World and still survive in its form. To me, these things are very interesting. I have much to learn. I have learned much since I came into contact in the past few thousand years and there is, indeed, much more to learn, for learning becomes a part of one's self when truly learned. When truly observed, these things do become part of one's self, functional, useful and purposeful. How can one prattle about things he does not understand? Yet many in your world and time are so doing. False prophets and teachers they are called by Christian dispensations.

Yet, perhaps I, too, at one time was guilty of these things. I sat in my Temple; I walked up and down in front of the great idols and chanted the mantrums and waved the incense, and called the people to worship. But that was a long time ago and now I have learned much better. Yet this is indeed only a form or an outgrowth of such various religious systems and cultisms which are practiced by those who do not understand.

Only one little bit of advice if I may interject; there is some note of caution that I would carry to you at this time; that it is best to approach the seat of learning, and when I say that, I mean when you call upon Him who is in contact, that this is done with purpose, it is done judiciously; it must not be promiscuously. He has much to give, much to teach, but He cannot throw what He has away. You are very worthy of learning all of these things, but in the future it must be observed that there must be a better time, a better purpose; that, whenever the thoughts seem to flare up, may I suggest somewhat more regular sessions wherein certain contacts can be attained and achieved without the necessity for Him to have to bob in and out as it were. (Yes, indeed.) It is something which is quite difficult but which He is quite accomplished at doing; yet there is just so much that can be done in one cycle. Too, it would be of great importance to you to have, as we do tonight, a manner and means in which these things could be perpetuated for others to enjoy them just as you are doing. If you need to, buy such suitable mechanisms which will give the greatest comfort to these dispensations. Remember, dear Sister, there are no limits to the contact this Mind can make. And a great deal of this depends, of course, upon how receptive you are, and what kind of a vehicle you, yourself, make to the outside world in these dispensations.

To Him, all those who are receptive, are worthy. He does not discriminate; yet He knows in the consciousness of the earth world how destructive these entities are. Am I speaking above the level? (No, I understand.) You have about one-half an inch of tape left? Did I get the correct amount of tape, Sister? (Yes.) I wish to show you that I do not see with the physical eye; to show you that someday the physical eye, the five senses, can be entirely supplanted by a higher form of intelligence. And that vibration, as it is concerned in such reactive processes as are associated with the five senses, is strictly something which we have learned to use with the five senses; but it does not necessarily prove that such things are nonexistent but perhaps five hundred or five million different kinds of senses. You will find that is true when you pass into the spiritual, just as the One who lives with you has found.

And in this proportion of contact through the mind with an infinite proportion of intelligent wave forms which you call the Spiritual Worlds, He is in a sense using five million senses, is he not? Five different points of contact with your different five senses means only a certain set of vibrations can be intercepted and relayed to the subconscious. With Him it means that five million or even more different forms of consciousness are intercepted and relayed to the surface of His mind. This is simple mathematics. You will find that it is true. And now I see that, as in the terminating point of all consciousness of your earth, the nine hundred feet of tape around the reel has almost exhausted my little stay. If I have added one note of harmony to your life, I am most grateful and I am your humble servant,

Lahamahasa.

(June 13, 1956):

A mental transmission received directly after the first of the Pulse of Creation books was completed and upon bringing a copy home, the Higher Ones related:

We, throughout the seven great Centers of Unarius, have been observing with you, these hours of fulfillment in the first step of the great dispensation of the seven books. We are not only those who have assisted materially in the transmissions or in other ways, but are a united concordance of the many millions of souls; nay, may I say billions, throughout not only Unarius, but in many other astral, cosmic and celestial worlds. Events of this nature which happen in the earth plane or terrestrial dimensions have a much greater effect in these various Spiritual Worlds than that which is immediately apparent in your own environment—something like the familiar skyrocket which starts with a small burst upon the ground but ends with a great display of beautiful pyrotechnics in the heavens above you.

We are all most grateful and joyous in sharing the joys and blessings of these hours and we know that, as all good things are like blessings and that blessings multiply, so now you may rest assured that all efforts shall be redoubled, concentrations of energies shall be fortified and intensified. We would like to commend both of you individually for your complete sincerity and dedication in helping in the fulfillment of certain evolutionary betterments which are coming to mankind in the future.

Before leaving we wish to especially compliment Ruth on her heroic and unstinting sacrifices, as she has shown in this day and hour, her complete dedication. She may rest assured all these things are well marked. For her experience this evening, we came to

123

her as Emissaries from the 33 Logoi with a token or emblem of our love and esteem which we placed around her neck. This is, of course, in the spiritual awareness and cannot be seen except clairvoyantly. It is a large, beautiful, golden, star-shaped medallion which has 33 points with small crystal balls at each point. These 33 points are sectionalized or divided in circular form, resting one upon the other and joined in the center upon a circle with a red ruby-like stone in the shape of a heart. It hangs about her throat suspended on a golden chain, coupled together with 33 links, each one joined by a small heart of purest crystal. It carries its own message of love and esteem.

As you know from all of the lessons, we do not believe in signs, portents, or cabalism, for only energy from the inward consciousness is intelligence. This is just a token of love and esteem that we have for you. It shall be around your neck to give you added strength and comfort, and will be there as long as you wish it so.

Do not be fearful or fretful, nor should you stress too emphatically the writings of any particular book, as this may tend to block something of the plan; but rest assured, all is well. Yes, I hear you asking who we are. May we say we are some of those who have been some small aid in the transmissions, and we shall be near in all your hours in your future so long as you so desire. We represent the many, but are momentarily, individually, Ishmael, Carmichael and Quetzalcoatl.

Mental Transmissions Via the Moderator to Ruth

Of Life Equation

(After "The Voice of Venus" book in manuscript form had been loaned to my lady friend, and who was unable to accept it as fact—she, having been the mother of Jesus in Jerusalem—it caused me grave concern. Thus, this reply from the inner planes from Master John, who was her polarity.)

You have been wondering why someone as high as "Mary" could not properly understand our transcript, or that I did not so influence or direct her, but in order to best understand how it is these things come about and that you may see more clearly, I will point out and describe to you more of the abstract equations of the individual soul progression. Part of this has already been given.

We may start at some individual's spiritual birth, and as it was similarized in the cycle of life equation, the individual began as an infinite concept. In this life cycle were the many multiple things which were all part of the wisdom of the Infinite God, and that as God is Infinite, He expresses both finitely and infinitely. Thus, in the life cycle, we see two diametrically opposed polarities or opposites, which so separate him into something like two poles. It is the oscillation and the various transferences of energy forms between these two poles which begin to construct for him his psychic body, or rather, two psychic bodies.

The individual will assume, in the course of many spiritual evolutions in each of these psychic bodies, two equally opposing concepts of such truths or structures as have entered into his psychic make up. I would point out that the two poles, as I have called them, are linked by the life cycle. This is, in essence, his soul or spirit. So it is that he goes through many spiritual cycles or reincarnations before he is ready to be initiated in this new phase of concept. Thus, he will have arrived at the point where he must reincarnate into some lower physical form in the earth plane dimensions. This he will do in a separate or divided state of consciousness, as he now becomes male or female.

This is the more enlarged concept of the biocentric or biune evolution as it is known on your earth plane. Up until the point where the individual had not yet embarked onto the earth plane, he had devoted his progression to (and in) the spiritual planes. These, as I have said, related him to such educational and constructive factors which enabled him to build two psychic bodies composed of two diametrically opposed concepts of these evolutions. I do not mean good or bad; I mean that these two psychic bodies assimilated intelligent wave forms within themselves by viewing and learning of God and of his great universal plan in different relationships. Like a little boy and girl who go to the circus and sit on opposite sides of the ring, they both see all that takes place, but just from different sides. Now this individual initiation or baptism will be into the lower earth plane orders. He may appear as male and female or he may appear as two individuals in the same community.

Now he and she may begin, through various evolutions—which have been described to you in other transmissions—as father and daughter. They may be two close friends or they may be brother and sister, but in all their evolutions and incarnations, they invariably seek themselves out and find a new relationship. So it is thus, that they are learning the supreme wisdom of God in these many incarnations.

126

Now it may happen that after several earth plane recurrences, these two individuals, or one biune, whatever you like to call them, will become somewhat attuned with the higher nature of God. They may thus obtain a new and greater understanding in their personal relationship with the great Universal Brotherhood of man, so they will be fired with the zeal of humanitarian constructiveness and thus begin their slow progress upward which will ultimately place them in a higher spiritual dimension.

Now it is so that when such an individual reaches a plane of consciousness which is comparable to some of the planes which have been described to you of the higher astral realms of Shamballa, (now called Unarius) that this individual wishes to express what he has learned for the benefit of the lower orders of earth people. He will then either separately reincarnate his psychic body into a physical relationship or attach himself into the newly conceived infant. This infant body thus becomes the instrument of expression.

Now it is obvious that the higher developed soul form could not function properly in a low or physical order. This he does through his psychic body which is living in the individual. You might say it is something like sitting in a phone booth talking to another one miles away, and that while you are sitting, you are conscious of your own world about you, just as the individual on the other end will be obeying or displaying such knowledge as you reflect to him through the wire. Thus it is with your friend, the teacher, Mary. As I have said at one time, she was on the earth, although this is not, in an absolute sense, quite so. As Mary, she is sitting in a phone booth. It might be like she is poised halfway between heaven and earth, but while she is thus situated, she is directing her psychic body and, indirectly, the movements of relationship of her physical. This, however, is done only in the spiritual sense. It relates only to the direction or the inspiration of word forms, energies and actions which are of the highest benefit to those about her, as her psychic body has become, through its many evolutions, a composite amalgamation of wave form energies which relate her not only to her true spiritual self but to her many earth lives. This is the reactionary or earth form body as it is expressed in the earth plane relationship.

Thus you can begin to realize, as it is with all mankind, and especially on the earth plane where there are separations of dimensions in soul consciousness, that some very spiritually intelligent people sometimes act extremely elemental when they are not in complete attunement with their true spiritual selves. It will also explain to you how the knowledge and wisdom is conveyed into the earth mind as it is learned from their higher astral world. A moment in attunement with the true self brings greatest wisdom and knowledge. As Kung Fu says, "A vessel will hold only its largest measure".

Your Brother John, (Master John)

An Open Letter

Today the peoples of the world are anxiously seeking and looking forward to world peace. This anxiety has been increased a million-fold when it is realized that modern science has placed in the hands of man, the potential power to completely destroy the world. History reveals that for thousands of years man has followed this same destructive course of evolution and that today he is no closer, yea, even farther away from peace than ever before.

Paradoxically, while he is thus anxiously looking to the leadership of his nation for this peace, he has not realized that he, himself, just as every individual, has a full measure of this peace and that if he is willing to make personal adjustments in his life, form new thought patterns, he can quite easily, over a period of time, acquire this much sought after peace of mind.

In order to properly understand what this peace of mind is and how it is thus developed, let us first transfer our introspection into a branch of science called psychosomatic medicine, and which deals directly with all functional attributes of human behavior. It is thus understood that each individual, in going through the first early years of childhood, suffers from day to day from various minor or major deflations of ego. Apparently upon growing to adulthood, this person has adjusted fairly well to the world about him; yet deep within the subconscious mind there is constant strife and turmoil against these various ego deflations. This strife always forms the backbone of the various patterns of conduct by which each individual conducts himself, even though he has succeeded in masquerading them to himself and apparently to others. Thus he has become his own betrayer and his worst enemy, and in his pattern of life, always finds various subconscious ways of fighting back against that hostile world of his childhood which made him suffer. Looking about, we can find people thus using these subconscious strife motivations in various ways. Their world has become a centralized, personal citadel. All people with whom they come in contact must bend themselves or be broken against the walls of these selfish, self-centered thought patterns.

When those subconscious strifes reach certain intensities, the individual then becomes more extroverted. He goes out of his way to force himself on others, thus subconsciously justifying his deflated, childish ego. And often, these people live in expensive houses; they acquire great wealth. Sometimes without talent or ability and with nothing other than sheer will, they become public figures.

In our final analysis, it can be said that the greater masses of population are much too subconsciously anxious for their own personal security to be concerned over their neighbors. Peace of mind cannot, nor shall it ever be acquired this way. Peace of mind means that we do not isolate ourselves completely from the various pressures and compromises which are associated in our daily intercourse with our fellowman, for the collective strifes and insecurities of each great nation become a great dragon which will ultimately fight other such dragons and thus destroy the world.

Acquisition of peace of mind is easy, provided one simple rule is followed: every individual should become conscious and mindful of the well-being of every other human being. He must immediately rectify any tendencies which are subconsciously extruded motivations in which he is still a small child fighting back against the world. In the words of the man of Galilee, "Do unto others as you would be done by." This is the simple, easy solution to not only world peace, but peace of mind with each individual.

The person who lives in a fine, palatial mansion, wears expensive clothes, has a yacht or drives costly imported motorcars is merely hoping to overcome his childish strifes by smothering them in an affluence of material possessions. Some keep noisy, destructive dogs, subconsciously transferring their hostility to these animals. Some have personal masochistic tendencies; others smoke, drink, or take drugs. They require frequent operations or may even commit suicide. Yes, even the woman who keeps a screaming peacock has subconsciously transferred her fighting, childish screams into the throat of this screaming bird. She takes sort of a masochistic pleasure in realizing that it is annoying her neighbors and tries to disguise this unkindliness by saying to herself, "I love them; they are so beautiful." If this woman wishes to really find something beautiful, let her look into the smiling eyes of her neighbor whom she has made happy by an act of kindness.

The way to true happiness lies not in the direction of that selfish pathway of domination and forcing our likes and dislikes on our neighbors, but individually and collectively, we will find happiness and peace only when we do unto others as we would have them do unto us.

The United Fraternal Brotherhood of the World

131

A Message from the Universal
Brotherhood on Prayer

Greetings, dear ones! Again, may I say it is wonderful to be with you, and through the medium of the tape recorder, to share with you certain concepts and principles which I know will be most valuable to you in your future evolutions.

In various works of Unarius, and particularly in certain articles in "Infinite Contact", assertions and comparisons are made regarding our present-day religions, particularly Christianity, which referred to them as pagan. While historically these assertions do not need defense, any serious individual with an open mind can ferret out in various existing documentaries, sufficient history regarding religion to verify the undeniable fact that all religions of today, including Christianity, are merely extractions and derivations from age-old concepts regarding various deistic configurations and subsequent religious structures constructed around them.

Other statements posed in these Unarius articles, referred to certain practices in these religions and particularly with the expression of prayer, that prayer has been labeled, along with other religious concepts, as likewise being pagan and barbaric in its content and in its psychological deference to the individual.

Now I realize that these statements may arouse a great hue and cry among those who are adherents to some Christian sect or denomination. In this respect, I am placing myself out on the limb, so to speak, and exposing myself to a lot of sharpshooters who would like to see me placed in some position whereby I could not deprecate from their false philosophy. In this respect, I am not making statements in my own defense, for none are needed. Like the historical aspects of religion, prayer as an expressionary denominator in such practices is to be as much shunned as should religion itself. I am not alone in these assertions; in fact, the so-called founder of the Christian religion, or at least the person to whom is attributed the motivating source of Christianity actually preached against religion and prayer, just as I am doing today.

A search through the New Testament and various other depictions of the life of Jesus prove conclusively that his sole aim and Mission was to tear down and destroy the various existing religious aspects of prayer. One of his teachings and commandments was (and I quote): "Ye shall not pray as does the heathen, in public places or on street corners, neither in temples nor synagogues, but thou shalt retire into the secret closet of thine own self and seek out the Father Within." Then, knowing the frailties of human indispositions as they are carried from one life to another, He realized all too well how almost impossible it was to change the inner nature or the reactionary self of the human without great preparation and without much practice—and without numerous lifetimes. So He gave them a prayer which is familiar to all Christians. However, this was done reluctantly and, as I said, only because of psychological deference to the frailties of human nature.

In an analytical sense and in making an analysis which could be said to be psychological or scientific in nature, let us dissect the reason or the reasons why prayer serves no useful purpose. It has never accomplished its intent; neither will it ever do so. There are perhaps millions of people who could and would become quite irritated about this statement. They would attempt to cite numerous incidents when their prayers seemed to be answered. It is quite logical and thoroughly probable, that in the law of averages, coincidental factors would show that many of these so-called answerings would have happened irrespective of prayer.

133

First, prayer is, in itself, a direct connotation to all past religious dispensations. The prayer the Christian utters to his Christian God is, in fact, in principle and in toto, the same as that being uttered by the black man in the jungle, the witch doctor, the voodoo priest, the Mohammedan in the far-off land of Arabia, or in any other countless dispensations which we may find about the surface of the earth. They are all invariably and basically the same in nature and principle. In this prayer, the individual so involved attempts to appease his god and, by whatever name he calls this god, appeasement in itself is quite ridiculous.

The prayerful person in his supplication quite often mentions many things which he would like this god to do for him, the reasons always basically being motivated from some selfish or self-inclined instinct of self-preservation or into furthering such various perspectives which he has in his immediate life. It has not occurred to this supplicant that he is trying to intimidate his god to whom he has given great power. He has usually, just as in all common religious prayers and deifications, given this god or gods the power of life and death over him. He has conceived in his god the power of creation, in creating plants, animals, worlds to live in, yes, even the heavens are all part of the power and of the creative energy of this supposed god. Yet with all this, this great god, supposed to be super-abundantly endowed with great intelligence and creative ability, still does not have enough foresight to anticipate the wants and needs of his abject supplicant!

A second psychological principle which can be considered most destructive in the concept of prayer is the dependency which grows within any person who has personified the ultimate in creative or mandatory elements in his own universal spectrum. He becomes reliant upon his prayers and to the god or gods for the various beneficent influences in his life. He therefore becomes mentally atrophied and cannot face life realistically. There is always a defense for various inane and subjective or negative attitudes and actions. He believes that with proper sacrifice and supplication that god or gods will forgive him for this sin; his moral perspectives have been weakened. Just as certain concepts are posed by a religious organization of this time in confessional, the priest has the power of eternal life, to the forgiveness of sins, etc., to any individual who places his soul at the feet of the priest. Such ideosophies and concepts are ridiculous, and if they were not so tragic in their consequences, would be actually humorous.

So long as any individual depends upon the pap and continues to suck upon it, he will not have either moral integrity or eternal life, for moral integrity, the strength of purpose, and the ultimate attainment of a creative function as an integral part in the scheme of infinity depends on how much ambition, inquisitiveness, purpose, dedication, faith to achieve, that any person has. Unless any person possesses all of these most necessary requirements, he will not attain either a fruitful material life or an eternal one. The two, in all respects, are synonymous to each other. The physical or material life is only one of the stages of development toward a preconceived or possible spiritual evolution which is devoid of materialistic aptitudes and attitudes.

To any person who prays and it makes no difference to what god or power he prays—he has automatically placed the power of mental introspection from his own self, into the hands of his deity. He has laid aside his birthright of personal integrity, of self-respect and moral development and given this false deification the power of life and death over him. This is indeed quite contrary to not only an intelligent equation of a progressive evolution, but is also quite contradictory to what could be supposed to be an infinite plan of progresssive evolution, as was conceived in the most perfect of all minds: the Infinite Intelligence.

Now when we enter into the concept of the Infinite Intelligence, we must lay aside all personal equations or suppositions which might involve various deifications. The idea that the Infinite Intelligence is some white-robed Santa Claus sitting out in the sky somewhere, with a big book of life sitting on his lap where he writes down our good and bad deeds, is a very childish and elemental concept, borne out of the miasmic past when deifications and religious forms took on various other expressions.

The Infinite Intelligence is, as can be correctly implied from the term, "infinite in nature"; that is, through great dimensions which Einstein called the fourth or fifth, etc., and which are actually certain planes of relationship, harmonically attuned to each other so they form sort of a constructive chord-like relationship to each other, we find a great regenerative, reconstructive flow of energy. This is, as has been described in the works of Unarius—assuming various interdimensional forms as vortices or vortexes—great whirlwinds, if you please, of radiant energy which are revolving within themselves in cyclic motions and which also imply centrifugal and centripetal motions, this process often in various relationships, culminating in such astronomical configurations as universes, galaxies and solar systems. Yes, even the suns and planets themselves can be considered apexes in these centrifugal cyclic motions. This is, of course, all far beyond the vision or the conception of the human mind. It must be remembered that the human mind as you now use it, is solely comprised of such elements as you have contacted in the past and in such basic experience quotients which you have extracted from them and which reside in the psychic anatomy in similar vortical patterns as I have just described to you. The motion, so far as the adjoining dimensions are concerned, is universal in nature and is perhaps the most comprehensible approach which the average human mind can have in understanding this great Infinite.

It cannot be subdivided into personal elements which can be said to be gods or godlike configurations. As a matter of fact, any individual who has ever attained a position in his evolution whereby he can conceive some portion of this great and vast Infinite can be said to possess many godlike proclivities and attributes. When a person arrives at such a dimension of introspection in his progressive evolution and in realizing the vastness of the Infinite, he is not, as is commonly supposed among the various races of mankind, an irrational or irascible individual, possessed of temperament and various other human frailties. Instead, he oscillates universally in a constructive pattern and does so in a manner in which he can define certain perspectives of this Infinity into well-ordered dimensions of transition.

This individual can also be correctly assumed to be beyond the point of bribery as it might be posed to some lesser earth individual who would be offering up to him or to his fellow humans in a similar plane of elevation, some passive element, or that he was, in some way, trying to conciliate this god or godlike person into some motivation best suited to serve the ends of this earth individual.

The highly developed godlike individual could see introspectively the progressive evolution which this earthman would have to take. He would also see any and all other factors involved in this transition. Time or place would be of no consequence and are secondary in their importance, the prime and most important consideration being in the general evolutionary aspects of this earthman. In viewing in such an introspective manner, this godlike individual would have a great, compassionate understanding for the poor earthman who was on his knees, trying to appease Him or intimidate Him into doing something for him, which was quite obvious he should do for himself. Perhaps at this point such situations could appear to be quite humorous if they were not so tragic, to the highly developed individual who knows quite well the destructive repercussive effect such continued negative expressions have upon the psychic anatomy of the earthman.

As has previously been stated, prayer is a very destructive element in anyone's life quite contrary to the general aspect which is now held by millions of people—because, as it has been stated, it detracts from personal integrity, gives a sense of dependency on unknown and unproven values. Also, it weakens the purpose of any human to acquire a better way.

So do not succumb to the rituals and habits of ages past that will only lead one down the proverbial garden path. Become intelligent and do as Jesus suggested—seek out the Father which is within each person, and in so doing, man can receive and experience the inner help and guidance so desired and, as He said, the Father shall reward ye openly.

Be Ye Born Again
(Of The Spirit)

Jesus said to Nicodemus, "No man can enter the Kingdom of Heaven save that he be born again." Jesus went on further to explain that what He meant was a Spiritual rebirth. To a greater portion of some 900 million orthodox Christians, this process of being born again means that at the end of some fancied millennium of time, someone blows a horn and they will again arise out of the ground and in bodies which have been somehow miraculously reconstituted, enter into some fabled "New City of Jerusalem" and walking on streets of gold, enter into a life of complete luxury and ease.

Such a story is not only fantastic and unreal, but belongs back in the primitive past and should have died with the semi-savage people who believed in such stories. Only in a realistic approach to life can we find the answer to spiritual rebirth which actually means through a large number of progressive cycles can any person evolve to a position which can be considered truly spiritual in nature and in this state he does not need a physical body in which to live; instead, he will have a different kind of body composed of refined energy quotients, polarized into a Superconsciousness, and that this developed energy body will enable him to live in a higher plane of consciousness and in worlds which are not formed from atomic constituents. This refining, polarizing process has been called reincarnation—another very poorly understood principle of life.

Those who have joined Unarius have found in these teachings a plausible, true to life and very scientific answer to all of these mysteries and misbeliefs. In general, it can be said that all those who have contacted this Center have been benefited. There have been those however, and fortunately a very few who have not seemed to recognize their opportunity or perhaps they had not reached that proper point in their evolution. There are others too, and fortunately a large majority who have found in Unarius the answer to all things. They have been "healed" of many physical and mental conditions and the whole world around them has changed. And there are some among these too who do not seem to know what it is all about and what is happening to them. These students feel a sense of loss or sadness; they feel at a disadvantage to cope with the ordinary transitions of everyday life. They no longer have the necessary physical strength they used to have; neither are they interested in the things which but a few months ago formed the greater part of their everyday life.

It is particularly in the interest of these students we therefore wish to point up some very obvious things which are happening to you. It must be remembered that way back, and long ago when you started your evolution as a human being, life began to be sustained as a personal entity from one life to the next by a number of physical reactions. This, of course, included a gradually widening horizon of mental perspectives. This life was sustained because the entity so concerned believed all these things were necessary.

However, through the ages of time and in the physical and spiritual worlds, this individual became gradually aware that life could be and was sustained in Higher Spiritual Worlds without a physical body and without the sins and iniquities of the physical life. He began therefore, quite naturally, to long for this higher form of spiritual life. In his various earth lives he would seek out various religions which seemed to offer him a better chance for survival after death. In a sense, this development or quickening process can be called a millennium, not of a thousand years however, but a thousand lifetimes. Then he will reach a certain point or junction in some earth life wherein he will actually pass a certain line of demarcation in this particular cycle wherein suddenly he is made aware—so it seems —of all of the things he has dreamed and longed for and somehow remembered as part of life in between lives in those Spiritual Worlds.

But the old self dies hard and in particular, those various configurations called the reactionary elements of the physical life. They still form, at least to some degree, part of the fabric of the psychic body called the subconscious, and which are still the sole remaining supports which enable the physical body to survive. In the former years and lifetimes the libido or drive in physical worlds was an activation for survival which stemmed in an oscillating process from these various subconscious centers. When the point of realization was, as it seemed, suddenly reached, a great metamorphosis occurred and an entire shift of balances of life was immediately instituted.

In other words, this person now was attempting to live from the Higher Centers in the psychic which were more directly affiliated with the Superconscious. Obviously then, this would and does cause repercussions. The physical body, still attuned in the atomic constituents of which it is composed, could not in its old rate of vibration, function as it had previously done. Various other physical functions and values also suffered misalignment and consequent loss of function and normal physical relationship was a result of this shift of balances.

Obviously too, the new life cannot begin until atomic constituents are attuned to higher rates of vibration, or even eventually eliminated. This and much more will be part of the future evolutionary processes through which this person shall pass in changing from the physical human to the spiritual being, and which is the exact process to which Jesus referred as being "born again".

Those students who have realized the great change but who still feel a sadness or a loss in their physical world and its various relationships to them, should indeed immediately begin to realize the full meaning of all of the implications which this great change has brought into their lives. They should realize that this is the long hoped for, the long sought after point of spiritual emancipation—this is the "Great Day", the day that the "Clarion Call of Truth" has sounded on the trumpet of life, and has called them from the dim and darkened regions of their past and physical lives, and that these great shafts of Light are reaching out like helping hands to guide them into their "New City of Jerusalem", but not a city with "golden streets" and idleness, but of many and beautiful Spiritual Worlds; countless thousands of years to be lived, not in idleness but as a life of learning, a life of achievement and a life filled with purpose. Each day, each epoch as it is lived, will bring countless millions of challenges which will test and try this person in all the materials and all the structures of this Higher Spiritual Way, and as these challenges are met and conquered, each one will give in return new measures of truth, attainment and realization.

So, dear one, feel not badly about the sense of loss in these material values, nor should you feel lonely in a world that seemingly has for you little company with which to travel. Do not, like Lot's wife, "look back" and feel reluctant to give up this past, for if you do, you will be drawn back into it by your very desire and you will be "turned to salt". Instead, turn consciousness to the Inner Self; soon you will find suitable company with which to travel, for about you in the great invisible worlds are countless millions of people, those who have sought, who have striven and who have attained.

142

Be joyous that you are now beginning to lose the mortal flesh and pit of clay in which it must live. Mark each day with new attunements, new relationships, for in this way you will indeed be striking off the many karmic chains which bind you to this earth. Be not fearful of the future but look forward with eager anticipation to this future, and this eagerness shall soon supplant the old primitive, bestial urge for survival.

While the future is Infinite, yet the future will hold for you only such portions of this Infinity which you can hold within your consciousness and make a part of your life. Like the potter who molds the clay, you then are the master of your destiny. This is the true purpose of the master plan for every human being, for in this way each human being can eventually become a unit cell of expression, an intelligent being through which the Infinite again expresses Infinity.

Freedom From the Material Pit of Clay

It can truly be said, that in the many thousands of years and in the many written and unwritten histories of the world, the movement of Unarius is unparalleled in the vastness and proportions, width and scope of this great undertaking, which represents the culminative efforts of countless thousands of people in somewhat more advanced strata of life to help their fellow man.

From the ranks of Initiates, Adepts, and Masters, literati, poets, artists, composers, philosophers and scientists, all in more or less advanced position to the Infinite, they have banded together in a common union and through such existing channels as a suitable outlet —at the present time, existing as it does in our channelship. While this in itself was a Herculean task, it represented but the first step, for it was realized in designing Unarius that its chief appeal would quite naturally be to those who were struggling in the throes of karmic indispositions.

This meant taking all of the fundamental wisdom in the universe, and much of which is not known to man, and tailoring it down to a simple, direct lesson course with a preparatory quickening into several books which, while they were chiefly documentary, also contained much of the basic wisdom which would enable the individual when reading these various texts, to couple them together with the various preconditioning elements which he had acquired in the Spiritual Worlds, and thus free himself from his pit of clay.

The third step is perhaps the most difficult of all— to find these comparatively few people who could thus be helped. Many of them are so ensnared within the tightly wound coils of various superstitions, dogmas, and subconscious reactions that they are unable to fully benefit from or justify the purpose of Unarius.

A similar psychological equivalent can be found in the common story of the small boy who was dragged off to the dentist by his mother to have an aching tooth pulled. Screaming and yelling, he will fight every inch of the way, even though he knows it is the only way he can get relief. This same subconscious inversion is a common practice among adults, as well as children, and the best that could be said about such reactions is that they are extremely self-destructive.

To find in the texts of Unarius, ageless and priceless wisdom put into a compounded and easily understood scientific form, and provable by all known physical laws, as well as its 'miracle'-working attributes, when properly used by those persons who have met the requirements, is in sharp and vivid contrast to the fortunately small number who, though mentally and physically ill, fight, scream, and kick in various adult ways against this life-saving wisdom—even after they have invested hard cash and time in their efforts to free themselves from their own iniquities.

The old saying, a man is his worst enemy, is most applicable here in these few instances. Many others, too, inadvertently find many ways and forms to cheat themselves from the great dividends which could be theirs. As Jesus said, we must seek the Father and the Kingdom as little children, which simply means that no person can journey into the Higher Worlds still ensnared in the bonds of his karmic indispositions. And so even many ardent and progressive students of Unarius find other ways to loop new coils about themselves in their daily life. They adhere to the same vicious cycle of thought patterns which have kept them, like bits of flotsam, swirling around in the chaotic maelstrom of the material worlds.

Cancellation of karma and the beginning of mastery,

comes about only when the individual becomes positively analytical in his daily life, and he can assume his own burden and the moral responsibility for having created his present life from the past. The evils of these past lives are like a thousand devils haunting the darkened corridors of his subconscious world, ready to spring out at every opportune moment and to fight viciously to keep the door closed against the Light.

If you are one of these unfortunates who is thus victimized by his past and has failed to gain suitable or expected results from Unarius, then it is high time for you to take a complete self-analysis. Start to uproot these mental bonds and brambles which impede your path. Pry open that door into the subconscious labyrinth; let the Light of Heaven permeate these dark reaches and the devils will vanish, and you, too, can join that happy throng of pilgrims who have, through the doorway of Unarius, started up that spiraling Pathway into the Stars.

Further Discussion on Canceling Out Past Negations

From time to time and as it is so indicated, certain basic and fundamental concepts must be re-entered and re-postulated to prevent a common and erroneous practice which may be unconsciously indulged in by the student which would delete the intent and purpose of the Unariun teachings. In the beginning these teachings were presented in the simplest form possible to avoid confusion in the minds of those who were beginning to try and master these great and completely evolutionary concepts, for indeed, they may seem entirely contradictory to all known and common practices in ordinary societies.

One of the more important and fundamental Unariun concepts is that which is contained in the principle of cancellation, and as so postulated, refers of course to the cancellation of past experiences from the present life or from any and all other past lifetimes in which the student may have lived. For it is very important that these negative experiences and their negative relationship in the present life, as a source of illness or disease or other retrograde aspects of life, must be cancelled out in order to facilitate a more advanced evolutionary progression; for life is impossible in the higher planes of life and in the Spiritual Worlds when any person would be so weighed down and fettered with the material world dogmas, creeds and their attendant experience quotients.

And so once again, as I have stated, in common usage there is also a grave danger of malpractice in which malpractice of these important principles would be subverted back into the more primitive aspects of witchcraft, sorcery or even superstitious dogma. It must be firmly stressed and emphasized at this time that since the introduction of these principles in the Unariun liturgies, this all-important subject of cancellation has been most adequately and thoroughly covered, yet it has reached my ears that in some instances certain students are attempting to use and validate this cancellation principle in their everyday lives in order to correct, as they believe, certain circumstances or to connect them with past life incidents in order to attain this most valuable and desired cancellation. And in material consciousness they have forgotten, or at least apparently so, the entire mechanics of what actually has to take place before such cancellation can occur.

Now, we must refer again to the basic elements of science. Let us remember our north and south magnetic poles. Two north poles repel each other, as do two south poles, whereas a north pole and a south pole attract. The reason for that is, as it has been explained for instance in the chapter of the electronic man, we must have certain polarities occur at certain intervals whenever wave forms of energy are so manifested, and as everything is at least in one relative sense an expression of energy, it must subsequently be then, a wave form of energy which is traveling, shall we say, forward in motion from its source as a succession of positive and negative impulses of energy.

Referring again to your past lifetimes as they are contained in your psychic anatomy, all that you have been for hundreds, yes, thousands of years, is radiating into your physical body, into your conscious mind, into all your cells, molecules and into the force fields of the atoms in your body. When this is done, your body, your life, your thought processes then assume the same patterns of expressive elements as they did in the long ago. Your body maintains the same physical aspects, the same metabolism, the same reconstructive processes as it did several hundred years ago, even though it is not the same body; the impulses which stem from your psychic anatomy have, in their expressive quotients, been reconstructed from the basic elements which have been introduced into the present form of your physical consciousness.

This introduction and the beginning of the new evolution of the physical anatomy, of course, started at the moment of conception, and from there on the inductive and impelling wave forms from your psychic anatomy controlled the growth and development of your physical body and your conscious mind up to the point you now occupy. There were of course many external forces which you had to reckon with; also there is between this lifetime and any given past lifetime a certain time differential which, when it is properly translated into third-dimensional terms, merely means that while these impelling or radiating wave forms from the psychic anatomy do control absolutely, yet due to the time differential, there can be certain variances when certain physical manifestations so occur. These variances or differences can be so genetically inclined, let us say, so that your present physical anatomy does, in some respects, resemble one or the other of your parents.

The scientist of today is attempting to translate these differences from the code material contained in the DNA molecule, not realizing that it is this molecule itself which is most primarily influenced by these motivating impulses which come from the psychic anatomy, bearing in mind of course the time differential. Another way to look at the time differential is to look at your house. It has square rooms and a roof. Basically and in principal it is much the same as the house you lived in five hundred years ago or perhaps a thousand years ago, yet there are differences. The modern plumbing and electricity could be part of these differences and it is in this way that we maintain, shall we say, a progressive evolution. Now, this time differential is most important to remember. Simultaneously, and in some instances, it may be so coded that, for evolution to be constructive, it must be forward in motion. The time differential in the appearance or the remanifestation of energy wave forms into material substance, (as I have discussed) in the formation of your body, means that this forward or progressive evolution can manifest itself. As was stated, it manifested in the form of plumbing and electricity in your house; it can also manifest itself in the differences of your physical anatomy. And here is the most important part of that principle which we are trying to adequately explain to you. As I have discussed Infinite Intelligence, the Source or the Fountainhead as the beginning of all things, and as it is the sum, substance and form of all things, it is in Itself complete. It is entirely unemotional and functions according to well-defined laws. These we have also discussed and we will find these laws expressed as positive and negative polarities in different wave forms.

Now, to be forward or progressive in evolution means that we are looking toward the Fountainhead or the Source, but more scientifically speaking, we are also looking forward to a new or more constructive form of whatever particular objectivism we are manifesting. To some degree this is being done in everyday life. People all over the country and in many places of the world as you find them are diligently trying to live better lives; they are trying to make themselves more affluent; they are trying to get more food or better clothing, yet they do all of these things in a reactionary way as does everyone else in their society. As yet, they have not determined the manifestation of all material forms and substances as a materialized form of consciousness of Infinite Intelligence. And this is very important because you or anyone cannot live in a higher plane of life until you learn very adequately how to manifest, shall we say, all forms of consciousness from the forward motion of your evolution or looking into the face of the Infinite. And when this is most properly done it will be in itself very complete, very adequate, and in essence then you, as a person who can so manifest a complete consciousness of the Infinite in a forward, progressive motion will be one of those most advanced, higher intellects which we have presented to you in these teachings and which, I might add incidentally, have been seen on different occasions by many of the students.

Now, getting back to our original topic. Naturally, all of this discussion has been most necessary to lead up to this point, for the principle of cancellation of the past is so entirely contained within the preface of what I have just talked about. To look back into your past life and to recognize some past incident from that life in some present form as it is manifested in your daily life is merely a conscious acknowledgement of that past life, but it does nothing to you or for you; it will not cancel out the negative influence of that life unless all other factors of this process are included. I have postulated or presented these things before but it is human nature to forget sometimes. In the pressure of the moment whatever the experience is, one may seem desirous to go along with the whole idea and he may sense the fascination of this past life experience so manifesting itself at the present moment in a somewhat different form, but it is still basically and essentially the same experience.

Yes, this is indeed true, not only for you, the Unariun student, but to mankind as a whole wherever we find him on this or any other planet; but you cannot expect any cancellation from a past lifetime experience by simple recognition in the conscious mind. The reason for this is that unless you include that other all-important factor of this recognition, this cancellation will not occur. Again, here are the reasons: this experience traveled into your consciousness from your psychic anatomy as a wave form, manifesting itself we shall say, as one wave form in a positive and negative fashion. Present consciousness in the cortical layers of your brain then reinstituted, to some degree, from out of the fabric of your present memory consciousness, a similar wave form which, as you can see, carried the same picture.

Now, these two wave forms originating from the same source at the same frequency and manifesting at the same time on the surface of the conscious mind, did absolutely nothing to themselves, like the two north poles which repelled each other. These two wave forms which were synonymous in nature and with positive and negative appearances so exactly timed, were to themselves benign; they were non-inductive and they did nothing. Of course there was some time differential. The time differential here was merely expressed more in a negative-positive pulse with the most recent past happening; that is, the past life happening was in respect more strongly negative because of the complexity of the wave form.

You must remember that as an experience in life it could not be simply a single wave form; it must be like the wave form which carries the television picture from your transmitter to your receiver—a very complex wave form. It must carry within itself thousands of other little wave forms, all of which go to make up the picture as it so reconstructs itself as an energy movement within the molecules and atoms and their force fields within the phosphor-compounds of each brain cell. It is in this reverse motion that memory is again reinstituted as a picture form because it is an exact facsimile, an exact counterpart, an exact motion as far as a positive and negative impulse is concerned with the exact past experience, and if this past experience comes from a previous lifetime the principle is exactly the same, except of course, as I have said, certain differences within the very complex structure of this wave form itself which would go to make up such attendant differences such as plumbing or wiring, or any other differences which could be realized within the present experience and that very similar and exacting experience in a past lifetime.

Now, however, suppose there was, looking forward into the future, this same exact wave form carrying within itself all the complex patterns and other wave forms necessary to recapture or to reinstitute, or even to recreate a picture within your mind and through the brain cells. This it could do if it is Infinite Intelligence, and you must remember that Infinite Intelligence has already carried within itself the exact facsimile, the exact picture and the exact counterpart of everything which you have ever done in your life, in your past lives and in any and all other appearances in which even portions of your psychic anatomy were so concerned or in which you so participated in your evolution up to the present moment.

So you see, even though you are not consciously aware of it, Infinite Intelligence has already radiated, I shall say, into your being or into your existence, everything that you are, and unconsciously and unknowingly you have, in every act and moment of your life, taken some small part of this great Infinity and materialized it in your daily life as consciousness. In doing so, in using this consciousness, you have in turn polarized it with the self-consciousness into one of the pulsating energy elements of your psychic anatomy.

Now let us suppose for a moment that there was a small portion of some of these radiated experiences or lived experiences which, shall I say, reversed themselves to a certain degree and re-radiated your experience of them back into the Infinite. This could be done if you recognized this particular experience as something which was inspirational or which came from the Infinite Intelligence, even though you called the Infinite Intelligence Jehovah or God or whatever you called Infinite Intelligence. If you were inspired at the time to recognize a great and superior force in your life which had expressed itself to you, you would by this recognition have begun to polarize in the Infinite a facsimile of your experience quotient. In other words, you would at that moment have begun or you would have added to the beginnings or the cell of what I have called your superconscious or higher self.

Now everyone on the earth, even the lowest savage, has a higher self; however, it is just a nucleus. He does not know what it is or that it is there in some cases. In other cases he recognizes it as part of what he calls his spirit or his soul, but it is all too vague for him to realize or to understand as I am discussing these things with you now. He does not know that his spirit or whatever he calls it, like everything else in Infinity in the great cosmogony and the microcosm as well as the macrocosm, as it is so termed, has a dual personality; that is, it must have a positive and a negative side or polarity.

The higher self represents the positive facsimile of the lower negative self. It is your problem as a student to recognize the positive polarity of your past lifetimes as they were instituted in your life in whatever positive quotients or equivalents they so contained in their experience values as polarizing agents with Infinity, wherein you began to construct or to add to your superconscious or your higher self. It is your most ultimate goal in the future to develop the proportions of this higher self to that point where it is the all-inclusive body of your existence. Such progress does not take place quickly nor in any one lifetime but through many incarnations, when you do not have a lower self to the degree which you have at this present moment. Oh, yes, even though you develop this higher self, it too will have a dual personality. It too will have a negative and a positive self and it will be your problem in that world and in that time, even though it is much more advanced than your present position, the relative values as they are so concerned will be just as drastic, just as all-impelling, all-encompassing, and with just as great a number of differentials as they are in your present life. In that Higher World the negative side of your psychic anatomy or your personality can—and it will—cause you sickness, retrograde motions back into the lower planes if you do not work with the principles. If you are not constantly facing, shall I say, Infinite Creation positively—perhaps even much more so and to a much higher degree than you can even imagine at the present time—then surely you will reverse your progressiveness and retrogress. You will be one of those fallen angels I have talked about.

All of this goes under the common heading of those very inviolate principles of progressive evolution: if we do not travel forward then we travel backward. And so, dear student, we must remember at all times this all-inclusive and very important part of this cancellation process. When our past lives in some moment come into the present plane of consciousness, they are reinstituted in some new form. This will not mean much in your personal existence; that is, you will not obtain the desired cancellation effect unless you remember the complete mechanics which are involved in this. You must also have a higher source of intelligence which can, at that exact moment, reflect or precipitate itself into the recreation of that experience and timed exactly right so that there will be cancellation. Now this impelling energy force must quite naturally be the exact facsimile of that past experience. It must be tuned

155

exactly to it, in other words, but it must be completely positive at all times. That is, as the lower negative wave form impels itself into consciousness, at the exact moment when the negative and impelling experience expresses itself in consciousness, there will be an equal opposing, impelling force experience coming from the other direction from Infinity but which will be positive in its polarity and therefore cancel out the negative experience. Now it won't destroy it because energy, as a wave form, is not destructible; it cannot be destroyed. The reason for this is that while you have changed the positive side of this negative energy wave form which came into your consciousness, you did not change its negative side, and its negative side can and does remain part of your psychic anatomy as it so polarizes itself in other portions of that psychic anatomy. So as you progress in evolution you will be able to tune in or remember that experience in a Higher World but it will not be able to reinstitute itself in your consciousness as a positive or dominating wave form or impelling force.

Now you may ask, where does this higher impelling wave form come from? I have said that it should come normally from the higher self. Well, if you had a higher self that was able to cancel out all of these incoming negative wave forms, whether from your past lifetimes or whether you picked them up on the street everyday from other people, if you had such a developed higher self you would not, quite likely, be living on this earth plane. You would be living in one of those heavenly places I have described, unless of course you were like me and you chose to come back and explain these principles to other people who did not know about them—in which case and under certain conditions is a subject for another discussion.

And so for the moment you can say that your higher self, while it is the beginning of your new psychic anatomy, and given a good opportunity with your dedication and your will to develop into a higher spiritual being, it still is not sufficiently developed to enter into these recapitulations or reformings of past lifetime experiences. It still cannot exert a strong enough polarity, a positive polarity, or an expression of polarity at the exact moment to obtain that desired consciousness. And you can prove this to yourself because you are still living in the world and you are very much concerned with your own personal problems in your own material plane of expression. So, granted now that your own higher self is still not sufficiently strong and powerful or that even if it were, you had not remembered at that moment when the experience transference or recapitulation had occurred to be intelligently inductive to turn this whole thing over, in a way, to the higher or superconscious self so that cancellation could occur. In other words, you merely manifested the experience in your present consciousness which was inspired, motivated and instigated from the subconscious or the lower self.

Thus, in the future it is imperative to remember that while placing, shall we say, the experience in a past lifetime or that we picked up a headache from someone off the street, we can place these things and discharge them very adequately and very quickly if we can participate and instigate in this participation all the elements and the mechanics necessary for such cancellation, for this depends primarily upon that all-intelligent higher consciousness, the higher self or your superconsciousness, and if it is not sufficiently developed, then you do the next best thing. And here is where the Unariun Brotherhood comes in, for while you are not able to do these things for yourself or that your present consciousness is still subconsciously motivated from the lower self, then you should not necessarily depend upon, but learn to work with the Higher

Unariun Brotherhood who is there with you constantly at every moment to help you in this process until some future time when you are much more able to do these things for yourself.

Recognition of these higher agencies such as the Unariun Brotherhood and remembering the principles which are entailed and which must occur will be of tremendous advantage to you. In remembering these things, these elements of cancellation, positive and negative polarities at the moment in which past life experiences occur, or in the cancellation of negative impingements such as headaches or other conditions in your daily life, then the remembrance in these things, even though you are not able to do them yourself independently, being conscious of them and seeing them work will, in time, enable you to develop the propensities of self-expression to the degree in which you will automatically incorporate them in every conscious moment. And that will be the time in your future when you will be living on one of the higher planes.

So, sufficient unto each day the evil thereof—to paraphrase a very important biblical translation. At the present time as a student, you should use all constructive means to attain an intelligent beginning in your evolution; begin to incorporate into your daily life these scientific principles of creation which I am discussing with you, for unless you do begin to learn to remember them and to try to begin to use them consciously, then your efforts in trying to make yourself a better person or to ascend to a Higher World will have had no effect; your efforts will all be in vain because it is mandatory, it is inescapable. To live in a Higher World means, as I have said quite frequently, a constant and never-ending conscious participation scientifically with Infinite Creation. And any time there is a letdown there is a downfall.

Therefore, to the future then. Whenever your past lifetimes so manifest themselves as experience quotients in your consciousness as relivings or if you take on certain negative experience quotients from the world around you, being conscious of them is very necessary and important, but do not fail to remember the scientific principles which are incorporated in the cancellation process, for unless these things are remembered in consciousness, unless you attempt to work with them and practice them, then all your efforts will come to naught. Instead, you will revert back into some of the more common practices of witchcraft, superstition and ideosophies which have been adapted, adopted and readapted by the numerous citizenry who have and are inhabiting this planet at the present time.

Yes, this world is a Tower of Babel, a confusion of many voices and many tongues, all stridently speaking to express themselves, believing in the superiority of their own position, not realizing that in their very cries for recognition, their efforts for expression, they are less than the smallest mote of dust in the face of Infinite Consciousness. For indeed this is so. While it is that Infinite Intelligence or Consciousness does express itself to some degree in every human and in every form of life, yes, and it regenerates itself from one generation unto the other, yet the vastness and the complexity and the infinity of this great Super Intelligence cannot be envisioned within the confines of some small mind who is so stridently trying to exercise his franchise of self-expression.

The dignity of our human society demands such a human individual expression; our democracy is based upon it, yet in the vast preponderance and encumbrance of the multitudes of laws and the derelictions, the vastness and complexities of the numerous expressionary elements have, in themselves, largely cancelled out and made invalid the democratic principles of self-expression. Therefore, as a whole the society of mankind as a tool or as a vehicle for personal advancement or the betterment of self cannot be used and should not be attempted to be used, but rather it should be done, as the Man of Galilee once said, that we must seek this self-betterment, this achievement and this self-attainment from within ourselves, making it an individual project wherein we can encompass to some degree the vast magnitude of Infinite Intelligence and, as He expressed it, as the Kingdom Within.

So, dear ones, remember those principles. Infinite Intelligence or Creation is very scientific. It functions absolutely and inviolately. It cannot be deviated and it cannot be bent to your will. You must lend yourself to it. If you do not, death as a retrogressive factor of your evolution will be inevitable. In the duality of this presentation, conversely by becoming a more integrated expressionary element of this Infinite Intelligence, so can this immortality be achieved, a greater immortality and a more expressive immortality. But in this immortality do not expect to float around on a pink cloud and play a harp.

Progression is also founded upon the ability of your own personal expression or the way in which you can express yourself—the degree in which you can express yourself. And you cannot express yourself creatively, intelligently or scientifically unless you know how. And if we wish to place things, let us do so by becoming conscious of these scientific and inviolate laws of positive and negative expression and in doing this we will be developing our personality of the future.

The wave form is cancelled out in a sense that it is so changed or reformed, reshaped or it is broken into harmonic constituents. Yes, it may even be regenerated into other wave forms which may also bear similar characteristics but which, in frequency, do not instigate those repercussive reactions in the physical and mental anatomy of the individual wherein there is a certain deleterious or negative reaction which can be considered to be sin, evil or disease.

It is a very difficult proposition, at best; to translate the aspects of fourth-dimensional abstract concepts into third-dimensional language—and more difficult for the third-dimensional mind so constituted in evolution from a material world and third-dimensional factors to conceive. At best I can only present to you the basic and most rudimentary aspects of these fourth dimensional concepts and they should be, in themselves, sufficient at least to start you upon a progressive evolution. It will be up to you, individually speaking, to each and every one who is in reach of my voice or who will read these lines, if he so wishes to survive and to become a participant in a constructive evolution, to begin to use these principles as I have presented them to him and without any attendant confusion or derelictions with what he considers his own opinions, his own desires and what he has learned in his third-dimensional life as being formed opinions or the biased contents of such opinions which would obstruct such a forward evolutionary movement of progression in his existence. So then, forward and onward into the future, progressively, intelligently and scientifically.

The Positive Regeneration

(More on Healing)

In presenting the teaching course of Unarius to the general public and through such commonly accepted channels of advertising media, inasmuch as Unarius presents certain factors and aspects on the basis of healing, all of these tend to attract a certain group of individuals who are escapists. This group includes not only those who are presently suffering with mental and physical aberrations but also another group wishing primarily to satisfy curiosity.

To those who are ill and are suffering goes our compassionate understanding; yet, as the prime purpose of Unarius is to teach the individual personal healing and prevention, nothing can be done for those who are shifting their own personal responsibility onto the shoulders of others. To the merely curious, we do not wish to waste our valuable time. We are not out for the so-called monetary gain but rather, to reach people who not only need healing but who are honest in their own personal evaluations and who have the necessary spiritual preconditioning and other requisites necessary to make Unarius successful with them.

The world is full of strife and suffering. It has always been so and will continue in much the same manner into the future, for in terms of evolution, the world represents to the average individual only a series of steps called lifetimes on that long ladder into eternity. And as these steps are near the bottom of his ladder, man is in this position, attempting in the many thousands of years, to evolve from the materialist, possessed with certain animal-like characteristics, into a higher form of life. Therefore, to the great masses of people

162

swarming this earth, these conditions and their material values are quite necessary to them, as they have been formed from out of their past evolutions, and until that day comes when they can look over the threshold into the great spiritual worlds and see their relationship to the Infinite Creator, they must remain on this same materialistic plane.

To heal all people promiscuously from their various sins and illnesses would do more harm than good and is contrary to the Infinite Plan, for it has been so conceived in this plan that the necessary wisdom which will help a person to evolve into a great Spiritual Being must be propounded from his mastery over the material worlds, as well as from the necessary knowledge which will help him develop the new spiritual form, wherein he will live in these higher dimensions. No one can do this for him, nor is it handed out on silver platters by some priesthood or some so-called spiritual healer, and until the day of awakening comes to any individual, he must remain the beast-like human, prowling about for survival on some material plane.

So, dear friend, ask yourself these questions and ask them honestly: can you recognize that the various conflicts and aberrations, presently with you, are products of your own making from past lifetimes? Are you willing to make every effort in study and constructive application in your own personal relationship towards a better future? Can you approach this new reconstructive process without biased convictions and strongly formed opinions? And lastly, but most importantly, can you conceive the vastness of the Infinite Cosmos and your evolutionary climb into the future, wherein you will find in this Infinite, that life will be a direct proportion of any part of this Infinite which you can so conceive?

Remember, also, that an honest constructive start into this Infinite through Unarius, means complete dedication of purpose, long and diligent study and application through countless years and lifetimes in the future, and from which course, should you ever renege, would mean spiritual oblivion.

It will also mean countless hours of compromise with the old self, for sometimes the old self dies hard. It will mean giving up what apparently seems to be, at the present moment, the very sustaining elements of life, for these things of the material world are compounded from fractions of reactionary principles and are actually like chains of steel which bind the individual to the earth.

When you can honestly answer yes to these questions, then you will find in Unarius, the doorway of the inner self through which you can step into the Kingdom of Heaven within and findeth the Father which dwelleth within. This Father, the Christ and/or your Superconsciousness, is one and the same thing, and joined in this principle of trinity, you will also be joined with the Infinite, wherein you will find your true spiritual life—your true purpose.

In The World But Not of It

Without a doubt, of all the classical utterances made by Jesus which is the most thoroughly misunderstood is his explanatory statement on spiritual status quo, "To be in the world but not of it" and followed by his explanation to Pilate at his infamous trial, that "His kingdom was not of this world".

The reason for this general misunderstanding and misconception is simple and obvious. The number of people who have lived in this world but have not been of it have been relatively few and only they, and they alone, know the full implications and meaning of such a life. It should also be obvious that such a person would be subjected to certain pressures, criticisms and judgments, yes, even persecutions which were thrust upon him by the reactionary nature of mankind whose primeval instinct reacts to destroy that which is not understood.

So far as my own personal life is concerned here in this twentieth century, there is ample and overwhelming evidence to prove that I am not of this world, even though I am living in it, and in the last decade, my wife Ruth has evolved into and taken her place in a synonymous parallel position with my own life and in an entirety of aspects, once again validates the Unariun Principle. Therefore, we, my wife and myself, occupy a unique and unduplicated position in this world of today and are therefore subject to all aforementioned criticisms, judgments and persecutions.

Fortunately, however, a Mission of such magnitude, and founded as it is in a Higher World which, as a society of advanced intellects and Master Minds, Super Beings, etc., such a Mission has been given the promise and the fulfillment of protection commensurate to its total intellectual expression and general propagation. Therefore, as earth world expressionists and as Moderators of the Unariun Mission, we again once more viably prove in many ways our affiliation with the Higher Worlds' organization—ways in which great intellectual power is projected into the world which, in its total capacities, adjusts, heals, helps and positively reorganizes a host of physical and mental conditions which are the generic byproduct of all peoples who inhabit this world.

The proposition to be in the world but not of it suggests and creates a multiplicity of compromises and adjustments which must be made by any person in this unique position. Such compromises generally disassociate this person from any and all of the normal socialities and the usual amenities which are, to some degree, shared by people of this world. The fact is, such apartness detracts and alleviates any strength which might be necessary to make a compromise which would involve abstinence from such societies. Such apartness too might appear contradictory to the general aspects of the Mission which is the teaching of insemination of all knowledge pertinent not only to humanity but to life and creation.

Any person or public would therefore be prone to make this criticism that any person in any apartness from this world would be neurotic and would be expressing his neurosis as a misanthropic withdrawal. Again, as intelligent power is projected through us (I, the Moderator, and my wife), it is therefore important that we maintain a practical and consistent attitude in our daily life, consistent to the general purposes of our Mission, that it serves primarily those who have had some preconditioning to the possibility of an evolutionary ascension into a Higher World through the doorway of interdimensional knowledge.

We do not promiscuously deploy ourselves by projecting energy into the general environmental climate. Moreover, such promiscuity would quickly deplete our physical and psychic energies and we would become ill, even a possible termination in death. Rather, it is that we project energies intelligently and through the inner portals of consciousness to all individuals and to those involved in certain environmental factors relevant to a constructive evolution. Also to avoid useless cancellations, correct timing according to cyclic terminations is most necessary—a function of millions of advanced personalities who are, through psychic channels, working with the peoples of this world and can determine correct deployments and projections of numerous types of energy suitably tailored to perform a specified task.

The proposition of apartness also yields other introspections. I, the Moderator, who am in reality a person and a part of the Higher World of the Unariun Organization, do not have the same physical and mental rapport with the physical earth world and this is also equally true with my wife, Ruth. We therefore do not find the reactionary satisfaction, the glorification and the wonderment of numerous earth world aspects as do those peoples who are indigenous to this planet and whose evolution is founded in all the earth world reactionary values which have sustained and propagated their lives from their beginnings.

No doubt, too, our psychic memories help in the general absolving from this world. From our lives in between lives in the Higher Worlds and from our sleep-state nightly excursions into these worlds, our memories are filled with the scintillating visions of these beautiful places I have described in my books.

So let it be known, and to set the record straight, that we (I, the Moderator, and Ruth) have chosen and adapted ourselves to our apartness. It is hardly a compromise for we see the world about us as a vast jungle

167

of human derelictions, depravities, injustices, revolts, destructiveness and sadisms of an infinite variety and we wish no part of them. We see not only the primary stages of primitive evolutions but also millions of people in more or less stages of retrogressive evolution. Ours is a Mission to ferret out and save those who have cried out for help, those who have recognized the possibility of personal salvation by learning how to live a better way. So it is from day to day, Ruth and I live in our apartness, confident in our Mission and with our affiliation with our Unariun Brothers in the Higher World, for we know that someday we shall join them in a complete entirety of union. Meanwhile in our world of apartness, to the multitudes of the world and through our auric radiations, through the intellect and understanding of our minds, there radiates great streamers of energy which penetrate into the deepest dungeons of human despair, shafts of Light which, when followed by the hopeless ones, lead them out of their carnal earth world. Yes, we know too that being in the world but not of it is being isolated from the synthetic superstructure of a material world, one which worships and idolizes the most perishable of human frailties, their superstitions and their false gods, their lusts, greeds and ambitions, their constant psychotic struggles to gain supersedence over their fellowman, their baseless struggle for attainment, for a super-ego.

Yet, within each one, we know there burns the same creative spark which engendered all creation, a spark which must be fanned into an all-consuming flame of life which will light up the dismal corridors of their earth world lives. And as this spark is fanned with the breath of knowledge and as the flame grows, so will they attain their stature in a Higher World. No, we do not really miss seeing X movies displaying nudism, eating out and enjoying gastronomic delights for-

168

tified with a host of chemical additives; we do not miss trips to Hawaii or Tokyo, etc., and staying temporarily in some glorified brothel called a hotel. We do not miss the stench of countless cigarettes, the boozy breath of those who would press about us in some nightclub; we do not miss the discotheque and their topless and bottomless waitresses. We do not miss the empty, hollow mockery of some great cathedral or church; we do not miss the fancied securities of some political system or belonging to groups, clubs, etc. We do not miss the applause and aplomb tendered by a multitude of hero worshippers and idolaters; we do not weep at funerals but weep only for those who refuse the most vital necessity of their lives. We do not cry for joy at some birth for we know the repetitious way evil is re-created from life to life. We do not pity the blind but only those who are so sightless they do not see the abundant promise of immortality which is ever about them. We do not condemn those who are false, lustful and perditious for their sins have already rendered judgment against them and the streams of a thousand worlds will be needed to wash them clean.

We do not falsely love, glorifying ourselves in adulation. We love on the premise that all has been created and is a part of the same source. We do not covet, for covetousness is a mire of quicksand and such as it is, mires the souls of those who falsely desire. Hatred is a blacksmith of the world who from the villainous passions of human nature, forges the strongest chains which bind hopeless souls in the deepest dungeons. There are a host of other similar factors and situations of the earth world which are gratefully missed. How else then could we be in the world but not of it? Let our desires be only for those who aspire to a better way, that we desire success for them in their future. We covet only the words which tell us of a great blessing.

How simple then is our life of apartness to logically interpret all ways, all aspects and all manifestations of life as a constructivism which openly displays creative principle. We bask in the serene, quiet complacency of our hillside home, almost totally surrounded by a wilderness of brush-covered slopes and hills. We listen to the birds and laugh at their squabbles over their rangeland and here within our hearts and minds is the peace, the strength and contentment which goes with the goodness of our work, and before us stretch infinite vistas of future worlds, lighted by the luminescent brilliance of creative love, an infinite understanding of the most finite participle of a created effigy.

So let our days and our apartness on earth be numbered. The Spring which brings the flowers, the golden leafy harvest of Fall, the abundance of all seasons which will, in their final day, be merged into immortality. For ours is a world lighted by Infinite Wisdom, powered by logic and molded into the sublime spiritual form by the ageless hands of time.

The proposition of being in the world but not of it has, in a certain way, already been well-covered and been publicized by science-fiction stories, science-fiction movies such as "The Day the Earth Stood Still", etc. In these presentations the theme or plot is similar: a visitor from another planet flying in a spaceship lands his craft on Earth where he immediately finds himself involved in a host of perilous dangers. Reacting in suspicion, fear and doubt, the earth people eventually try to destroy him. He may also meet death from a host of earth world germs and viruses from which he has no immunity, as it is assumed that he came from an advanced civilization from a distant planet where diseases had long ago been eliminated. These conditions which he would meet here on Earth would be vastly multiplied and augmented were he to appear in his physical body which was different to that of the earth people. It would make little difference that he came on some beneficent Mission to give the earth people the advantage of his advanced science and knowledge; either he would be forced to flee or to use some advanced weapon to destroy the earth people. Now all of this has been circumvented by the Unariun Brotherhood living in their Higher World. They have chosen to expedite the logical and intelligent way to bring the Unariun Mission to the earth people to impart their knowledge, wisdom and great healing power to the earth people, and especially to those who had in some way previously been partially conditioned or acclimated to this advanced science and the way of life which is lived when some mastery is attained. This logical way was through the doorway of reincarnation; an earth person with a physical body is born into the world.

This body, however, is motivated and activated by a psychic anatomy which has been especially prepared and put together from parts of psychic anatomies which motivated bodies lived in previous lifetimes and which were people actively associated with the Unariun Mission. Specific cases in point are Akhenaton and Jesus. Here in the twentieth century the same condition is true with the Moderator. The psychic anatomy of his wife Ruth has also been given great preparation in the Higher Worlds of Unarius. This is most necessary. All rates of vibration which are radiated and expressed from and by her must be compatible and harmonious to those of the Moderator, otherwise He would suffer a constant depletion of psychic and physical energies. As a matter of fact, Ruth has been the only person on earth whom He has contacted who did not deplete him and this was one way in which He knew she was the one He was looking for, and his many years of searching were over.

It must also be borne in mind the complex vibrating

structures of the physical body, the mental consciousness, the psychic anatomy, etc., of the Moderator are very greatly different and on a higher level than those of all other earth people. This higher rate of vibration is a most necessary condition, one which makes possible the channel of communication and deployment through the Moderator. The Unariun Brotherhood would not be able to bring their wisdom, intelligence, healing powers, etc., through ordinary earth people who were totally involved in their material lives.

It can now be much more easily understood as it applies to Akhenaton, Jesus and to the Moderator and Ruth the proposition which completely separates physically, mentally and spiritually or psychically these people from the material world which surrounds them. Their physical bodies are a most necessary disguise. To all appearances, manners and ways they look like ordinary earth people. They dress, eat and otherwise conduct themselves as do earth people. However, here the similarity ends. At any given moment in their lives here on earth, they are totally involved with the Unariun Mission. They are radiating wonderful energizing power into the world about them—energy and powers of many different kinds—all tailored by the Master Minds of the Unariun Brotherhood to meet certain specific conditions and for certain specific persons.

Also in line with all higher rates of vibration implied in this condition is a very great supersensitivity. The five senses are acutely conscious of odors, lights, light-waves, sounds, et cetera, far beyond the threshold of ordinary people. On several occasions, competent doctors have found the five senses of the Moderator to be from 22 to 33% above normal. Also most puzzling to the medics, He still maintains the blood pressure of a young athlete even though He is in his mid-sixties. Most recently his eyesight was found to be 20/20 and was most recently corrected from a stigmatic variation which had been with him for thirty years.

Since his birth, the Moderator has walked the earth projecting a tremendous auric power. It was this power, plus his genius-like precocity which set him apart from the world. Through his growing up boyhood world, he was subjected to an apartness which He could never bridge in whatever way He tried. To his fellow students at school. or to the few boys He associated with, as well as family members, He was to them, always somehow different. There was something strange and unknown about him—something which they could not quite understand and as might be expected, He suffered from a constant stream of indignities.

Many years before the beginning of the Unariun Mission and at the beginning of World War II, the Moderator's way of life drastically changed. The auric radiation increased as did his psychic perceptiveness, and his ability to see beyond the range of normal vision became a legend in the area where He lived. Then in 1954, there was one more step-up and acceleration and from there on, living as He did in close contact and communion with the Higher Worlds of Unarius, He has steadily advanced all rates of vibration and sensitivities.

The same conditions of advancement, psychic and spiritual metamorphosis have been the course of life with Ruth. She now is on a plateau equaled only by a few people who have lived on this earth.

Perhaps a few incidents would help point up what it means to walk this way of life. In the normal daily intercourse with the outside world, buying food, supplies, mailing service, clothing, going into different stores, contacting people etc., presents problems which are, in certain aspects, a bit humorous. When any persons are contacted, they are immediately engulfed with an overwhelming wave of power which causes them to react and do strange and incomprehensible things such as, making the wrong change, a sudden inability to figure, taking articles they have placed in bags and putting them back on shelves. They usually become extremely hot; the air-conditioner normally blowing cold air is now blowing hot; they may become red-faced, start gasping for breath, fan themselves and remark how hot the weather has suddenly turned. In a few instances, some persons have become antagonistic, even threatening to throw the Moderator out of the store, and for no apparent reason.

Even as He drives his car, there have been instances when those driving other cars in some proximity to him have exhibited strange and erratic behavior. Eating in cafes is usually a problem; waitresses get flustered, forget to put their order in or get it wrong. They may avoid the table as long as possible. In attending a movie, all persons who sit close to the Moderator do so for only a few moments then get up and move, remarking how hot it is here and eventually the theater is all filled except an empty ring of seats around him.* These are but a few of the many compromises and situations which must be properly met and dealt with by anyone from another and Higher World even though he may look and act like other people.

(*These outings took place some fifteen years ago.)

No matter how perfect the disguise, the difference and apartness is always with him and always to those He contacts there is a flow of power to them, a wonderful healing and adjusting power which leaves them with a giant step forward in their evolution.

In the past years Ruth also has rapidly advanced in her ability to radiate power and all conditions so described are equally true with her. Likewise, there are several students who have, since beginning with Unarius, begun to exhibit the same radiating power, proving how a polarization takes place with these students, made possible by previous preparation and conditioning.

Thus it is that we, the people of another world, walk the perilous pathway of the world, a thin line which constantly differentiates and dictates what we may or may not do, or that even if we would like to do, we know that the attainment would not match the effort; it would only be a flatulent accomplishment without the customary satisfactions enjoyed by the earthman.

To visit Rome, Paris, Athens, Honolulu, etc., would only be a hard accomplishment, one which we would no more enjoy than we would visiting the neighborhood park. We could only say we'd been there and as there are so few of us, we'd have no one to say it to! Personally, as far as my own experiences are concerned, I would not recommend it to anyone.

For the earth people, mysticism has always played a large role in their regard for such a person. The Christian religion was founded presumably upon the Mission of Jesus, yet there was no mysticism for Jesus, only hard work, walking many miles through the desert heat, getting handouts from some sympathetic villager, enduring insults and indignities and the unwanted adulation of multitudes, then finally an ignominious death brought about by those He tried to help.

So it was with Akhenaton and so it would be with Ruth and myself, the Moderator, had it not been that we have so adequately planned a great wall of protection about us; even our lives, despite the radiating power, are a protective device which, together with the laws of the land which protect us as they do others, we live in comparative peace, free from the danger of crucifixion which we would endure were it otherwise.

No, we are not misanthropic; we do not hate the world, the people in it, or the establishment. We accept all perspectives with a rationale which is based upon infinite perspectives and these perspectives place the world as it most properly occupies—a place in the astral world, a place of the beginning and also an end. Should those who are born into this material world or any other such world not succeed in continuing their evolution beyond its third-dimensional boundaries, then inevitably they will recede on the backward wave of their cycle.

We, the Unariuns, maintain our Mission in this world for the sole purpose of helping those who earnestly desire and have dedicated themselves to the proposition of going beyond this world to attain a long-dreamed-of immortality and the practical realization of life beyond the graveyard of this world. In our files are the handwritten testimonials of many thousands who have been so helped and there are other uncounted thousands who have likewise been helped but have not yet testified. Regardless, at the time of their passing from this world they will, after crossing the river Styx —the river of death—find a strong and willing hand to help them from their frail bark of desires onto the shores of their new Spiritual World.

Beware Ye In the Latter Days

Two thousand years ago a certain man uttered this most famous and prophetic prophecy, "In the latter days, beware ye of false prophets and teachers, those who dress in sheep's clothing but are as ravening wolves." Today in this twentieth century this same man has, through reincarnation, lived to see more than a full and complete fulfillment of this prophecy. There is on every hand the modern version of the Tower of Babel, the cacophony of a countless multitude of voices, false prophets and teachers, the mendicants and purveyors all claiming to be and to have various merits which they are extolling and attempting, for a price, to sell to a gullible public.

Politicos go from town to town raising their voices in protest against each other, each one extolling and promising what he will do should he gain his office. There are many and countless claims arising from different pulpits, religious groups, cultists, quacks, et cetera. Here in Southern California there is a tremendous gathering of these false prophets, a Mecca of sorts, due partially to the benign climate but particularly to a kind of social structure which is most affluent in nature. Wealthy or semi-wealthy people retired in great abundance, live throughout the communities of the Southland and here these false prophets and teachers, quacks and purveyors have found a lush field for the deployment of their wares and their nefarious practices.

It might also be noted that some attempts by different governmental agencies are made to curtail these individuals, yet little success has rewarded the efforts of these different governmental agencies. As of today, statistics reveal that some 500 million dollars a year in this Continental United States is being given into the coffers and pockets of false quacks who are selling cancer cures, cures for arthritis, cures for this, that and the other; even dandruff cures, wart cures and cures for hangnails and bunions.

Ralph Nader and his organization of Nader's Raiders are, to some extent, partially effective in suppressing certain activities, or rather, the lack of certain activities among the manufacturing communities of this country to those who manufacture vehicles or those who process foods and so on. But as of now, no one has come forth to make a concerted and worthwhile effort against the practitioners of certain false doctrines which are expounded under the cloak of religious freedom as it was set forth in the First Amendment of the Bill of Rights in the Constitution of the United States.

In particular, and as a classical example of the different purveyors of certain religious or metaphysical or mind science tripe and trivia which is being peddled in the United States and in particular here in Southern California, I am going to describe the activities of a certain, shall I call him, practitioner who has, through some twenty-five or thirty years of activity here in Southern California, succeeded in making for himself a rather substantial position in the materialistic field of his life. This practitioner lives on a hilltop in a palatial mansion; he has a butler and a chauffeur, many cars and lives the highest epitome of affluent life. Every day there is, in a large metropolitan newspaper, an advertisement which carries his picture, along with statements and claims which he makes in his practices: "Attract wealth to you through personal magnetism . . . how to develop your magnetic magical powers . . . how to do this, that and the other through the magical dynamics of personal magnetism, even to attract love and other things to you through this magical personal magnetism." And so far as he is concerned, it quite evidently has worked. Whether or not it works with those who have listened to him in the theaters which he leases and where he sits upon the platform while he speaks on what or how to do these things, is very doubtful. I have listened to this man on at least one occasion and I have come away with the same feeling which I have when I read his advertisements in the daily newspaper. I am wondering how many thousands of people he has succeeded in stalemating or deviating from a true constructive evolutionary pathway.

It is quite evident to a man who has had some training and some knowledge in the field of human nature that all of which he speaks and tries to teach borders very closely on the cultistic witchcraft which flourishes on the back-reaches of this planet. There is little to differentiate between the practice of trying to concentrate on the materialization through your mind powers to acquire an automobile, a better house to live in or any other such sundry physical, material acquisition except perhaps there is missing the witch's mask, the usual voodoo smoke and incantations and perhaps the amulet to hang around your neck.

Here, at this time and place, there is just a simple expression that if the person desires to do so, by simply thinking about it and concentrating on it that it will eventually appear. Now, such utter nonsense and tripe, is of course, completely fraudulent in nature. As I have stated before, there is nothing in the human mechanism, physical or otherwise, which could perform such a feat. There is, of course, the thought which is carried; a person can, shall I say, concentrate his daily activities into a certain direction and it may in a certain time, if he is constructive enough in those endeavors and puts forth the necessary effort, he may eventually acquire a car or whatever else he does so acquire.

The fortunes of capitalists which have flourished at this time and in this century are built upon the premise of concentrating their efforts, their lives into the direction of acquiring riches. This is, in itself, a certain type of constructivism, I shall call it, which does obtain certain results. But certainly there is no magic about it. Certainly the capitalist, whoever he may be, has not acquired his empire, his millions, by sitting and concentrating. He went out and he worked for it. He did a lot of things which, quite possibly, an honest man would not do, yet he did acquire the wealth, not through magic but by using all of the numerous devices along with great personal effort to acquire this wealth.

Moreover, it is quite beyond a question of doubt that in these efforts the capitalist, whoever he was, had the help of certain astral agencies; also, that he had been a reincarnated person who was, in a former lifetime, a person of wealth and so quite naturally had a certain acumen to pursue what he did so assiduously in other lifetimes—to accumulate wealth. And so the acquisition of wealth was not as great a problem as it would have been to the average ordinary person—a person who might have sat in the audience of this practitioner and listened to his dissertation on "the magic of mind concentration" and all of the other google-de-gook which these purveyors are constantly interjecting into the minds of their rather blissfully ignorant subjects.

It is also beyond a shadow of doubt that many of these people who come to listen to the trivia expounded by these mind practitioners are, in some way or another, immersed in some neurosis. There has been a great frustration, a defeat, a loneliness or some other traumatic psychism which is working in their present lifetime and is making life quite uncomfortable for them; or they may simply be attracted blandly enough to the banner of this circus barker, this side-show spieler—which he really is—in order that he might in some way financially incur a greater abundance of material wealth.

There are other prospectives, too, which present themselves in this discussion. I am thinking of this and other practitioners and of their numerous subjects which they have attracted to themselves; what will life be like for them when they leave their physical bodies? They have spent all of their energies listening to this false purveyance or they may have spent their energies in attempting to follow these false practices. The practitioner himself has centralized his entire existence into the acquisition of material wealth. There is no materialism in the fourth dimension or that spirit world, as it is called by some people. Any person who leads a normal and an abundant life in the hereafter must do so by virtue of his own mental capacities, his ability to understand the general fourth-dimensional Infinite. The infinite macrocosm is not a personal mendicant which purveys any particular participle of affluence to any person, either that this person should pray or otherwise attempt to intimidate the Infinite.

As I have stated before, the Infinite is absolute; it is law, order and harmony and functions equally for the smallest atom to the most advanced of the species of Homosapiens. There are no exceptions, religious, cultistic or otherwise, and if one hopes to aspire in the trajectory of a constructive evolutionary pathway, this must be done, not on any theories which seemingly incorporate certain magical formulas, but rather upon certain fundamental and basic understandings of what creation is and how it functions.

I doubt at this moment that this practitioner whom I have discussed knows what a sine wave is. I also doubt whether he could go into the mechanics of the human mind or that he possesses even the most rudimentary fundamentals of psychology. While he has made claims that he has journeyed to Tibet and that he has studied in different esoterical schools such as yoga, zen and others, yet in my research I found nothing in any of these oriental theosophical schools which dealt specifically with the construction of the macrocosm or the microcosm. These were all theorems, placebos, or the flatulent statements of something that existed—and we must believe that they existed—and this is a great flatulency of intellectual expansion.

No person can develop the necessary perspectives and all the incumbent requisites which are required to properly relate him to a constructive evolution without knowledge of the electrophysics incorporated in the construction, expansion, the propagation and the constituents of what the Infinite is, no more so than you could develop the necessary memory, the knowledge which would enable you to go forth on some journey to a distant city or around the world and then return to your own domicile or that you would be able to carry on any of the other necessary requisites that you might. These are only temporal attributes which you have developed through the concourse of your evolution and the furtherance of this evolution will concurrently and without any other malfunctions, carry you into the Infinite, provided you were completely objective, you were diligent and you were unrelenting in your pursuance of this evolution.

In all foregoing transcripts which I have dictated in preceding chapters of these texts, I have very specifically noted all of the mechanics involved in the transmissions of various mental faculties—paranormal or normal—those which are adjacent to and incorporated in your daily life as well as those which must be entered into, fully understood and practiced in your afterlife. The laws or principles which govern the transmissions of these functions as they concern your own personal self are inviolate. I do not make threats or

182

demands. I am merely making a common statement of fact. Either you do believe or you don't. If you do, fine; if you don't, well, then you are your own judge, your jury and become your executioner.

I am here for a specific purpose, one of which is if I can, to waylay all of these sinister and evidently quite materialistic motivations which engender the purveyances of these malpractices which we find about us in our daily lives, even in the practices of the ethical sciences such as they are being currently practiced in the medical fields. It has been a common admission made by doctors that some seventy-five or eighty percent of the patients who visit them would normally and without visits to the doctor's office get well by themselves; ten percent can be helped and ten percent will die through one means or another. Now seventy-five or eighty percent who cannot be helped, and only ten percent who can, is not a very good average. Perhaps in time this ten percent can be increased to fifteen or twenty-five, yet if this is so, then our society—such as it is and compounded of such heterogeneous expressionists who relate themselves to the fulfillment of this Biblical prophecy of false prophets and teachers—must be more fully purged of these elements.

While we are on the subject, we might enter into that field of public exploitation which is also protected by our Constitution and the Bill of Rights; religion. There is without a doubt, the greatest fraud which has ever been perpetrated on the human race; yet it could be said that strangely enough, by and large, the human race permitted this perpetration of fraud; he victimized himself, in fact, he actually demanded that he be so victimized and exploited. Through ignorance, superstitious beliefs, the supernatural, et cetera, all were taken advantage of by certain elements in different societies that built up an incumbent priesthood which lived off the fat of the land, so to speak, without work and without any of the other necessary expedients of life, literally had people coming to them, dictating their lives and their moralities. Yes, religion as it is today and has in the past, dictates their sex lives and it occupied the foremost position in the great frauds which have victimized the populations of the world. It continues today; it has continued from the beginning of mankind upon this planet Earth and until man throws off the shackles of ignorant superstition, religion and all its attendant offshoots of such expressionary cultisms, of mind sciences, metaphysics, et cetera, it will continue to flourish. As it is often said, "There should be a law against it." Yet the precinct of our Constitution cannot be altered or changed and would have to be so done in order to bring about effective legislative action and law enforcement, etc., which would curtail the activities of these false prophets and teachers.

Generally speaking and looking into the future as it concerns the cyclic movements of planets, populations of planets and their levels of expression, etc., and making some prognosis of this future, it is doubtful whether this Earth will attain any plus so far as the values which I have discussed in the numerous Unariun liturgies. The earth plane as a whole revolves in a position in the dimensional microcosm or, in infinity, the macrocosm, which relegates it to a certain particular level of life. Should this level of life exceed a certain plus or minus, then such life would perish, for the constituents of an Earth life are based upon reactionary elements such as the law of survival—the survival of the fittest. The emotional content of a person's life which either feeds or deflates the ego is the all-important and the supreme edifice of every person's life irrespective of any beliefs or practices which he may express to his fellowman. The ego is all-supreme and is all important; individually speaking, there can be no alternatives.

Even the documentaries which have instigated and propagated this nation as a whole—the Declaration of Independence and the Constitution—idealistic as they are in nature, hint, in fact in a broad sense, include a Utopian civilization, a long-dreamed-of Utopia by mankind in general, wherein all of the attributes, the condiments and necessities of life are immediately available, that there is no poverty, no sickness, no insecurity; in fact, there is no anything. Such a Utopian concept is entirely sterile; it does not belong in a logical mind or in a thinking process of any human being; yet striven for, it is the fulfillment and attainment of a so-called Christian life in his salvation and his resurrection, for it is Utopia in the New City of Jerusalem which he hopes to attain. How stupid!

The full purpose and cause of life is contained in the countless life experiences which relate a person to his own individual expressions whether they are constructive or destructive, whether they are good or whether they are evil. Within the circumference of these expressionary elements it is mandated that his intellect shall be developed to the point where he shall become selective, that he become knowledgeable of the factors involved in these expressionary elements, for these elements are immediately available in every perspective of his life, they are paramount in every inductive process.

Every life experience is based upon the premise of his selectivity according to his own emotional values and should he fail to make the proper selection, the ego will eventually be deflated and there must be a concession of what is right and what is wrong. This principally is the underlying factor in the failure of our present civilization. While schooling is conducted along in the importance of attaining the three "R"s—reading, writing and arithmetic—there is little or no emphasis on the psychological values of life as to what constitutes moral responsibility to every person, that he may falsely interpret what he sees about him in his daily life, that he cannot attain what he believes to be the opulence of wealth or that he may become jealous of his neighbor's success—inductive processes which sooner or later will create a neurosis should these things not be adequately discharged, rectified or otherwise compromised. For such daily experiences where a person is constantly subjected to such deflations and flatulencies, a neurosis is sure to develop and a neurotic mind is the beginning of a sick mind and especially so should this neurotic mind be unable to recognize objectively the cause, source and the therapy necessary to correct this maladjustment.

Thus our civilization, such as it may be called per se, as it is now, seems to be rushing toward a precipice, a civilization which at present tolerates indecencies which are unspeakable in nature. Art forms of all kinds have been degraded and vilified; there is a constant vilification of all of the attributes, the social structures and the common expressions of life. Yes, even sex is openly exploited and displayed upon a well-lighted stage in all its intimacies and for money—a situation which is tolerated, a situation where there is no law to prevent such indecent exhibitionisms.

Yes, and there are murders and rapes, the crime rate is constantly increasing much faster than does the population and one can only wonder how long it will be before every person in this country will be a criminal of some sort. Yes, and indeed he is, for the tolerance of crime is submission to it, is an acceptance, and vicariously every person then becomes a criminal if he is living in a criminal atmosphere.

Thus once again of the philosophies, in the speculations which will arise from these philosophies, we can only make a forward prognosis into the future to say that in the evolutionary course for man on planet

Earth, that he has raped this planet, that he has vilified his existence in all forms and factors, that he has exceeded the length of his own sword of wisdom, that he has perpetrated a great crime against himself. Yes, indeed, these are "The Latter Days" and the fulfillment of that prophecy has exceeded all expectations!

*   *   *   *   *

Part Two

Our total objectivism about the Biblical prophecy of the latter days and the false prophets and teachers would not be complete or final unless we made one last observation—a most obvious observation; that these individuals, these false teachers and prophets, the purveyors of self-hypnotic doctrines, the magic elixirs, the panaceas, the cure-all salves of the medicine-men of the modern twentieth century who have all adopted and adapted new forms of expression—all of these mendicants have their own particular way of communication, they have their own way of attracting the gullible public to them and to extract from them any and all of the affluences of life. And while they are thus purveying their false witchcraft, their dopes, their palliatives, their see-alls and their know-alls, they have one hand in the pockets of these poor souls who have come blithely to them, either with curiosity or with higher expectancy that in some way or another their problems would all be dissolved by these magic elixirs and panaceas.

It is quite obvious that in these systems of communications whereby these attractions are set up, advertising media of different sorts is used, such as the daily newspapers, different magazines which cater to this particular kind of tripe, and in all of these advertisements there is a common note which, like a bell that is so often struck, it is repetitious in nature, they are extolling the virtues of their system or their cure-all. Yet it is quite evident and very clearly implied that they possess all of the secrets, the knowledge, wisdom and truth, which are automatically incumbent with their particular cure-all or magic; yet they are forced, as it were, to constantly exploit, to sell themselves, to use communication or to otherwise shall I say, labor in their particular expressionary field for the means of survival!

Yes, it may have been that they may have acquired some wealth, some position of affluency; they may have cars and chauffeurs, palatial mansions, or they may be some humble spiritualist (apparently humble, at least), who is living in a back street in a small, dirty, dilapidated house and who advertises in a small section of the daily newspaper under the personals column that she sees all, she knows all and she hears all, and can give you the answers to all of your problems; or that person who takes a quarter-page advertisement in the same newspaper who is, in other words, telling you of the magic in your own mind, the magic which you can develop by simply sitting down and saying that it is so and concentrating on it. Either one way or another, these people are quite obviously unable to fulfill what it is they advertise and claim they will teach others to do, for if these particular purveyors, these mendicants could apparently practice all which they advertise, then the necessity of the advertisement, the deployment and the exploitation of their supposed teaching ability and the powers of their cures would be expressed in their own particular dimension!

Quite clearly then, there is something else indicated, a rather sinister indication which points a guilty finger toward some of the subconscious neurotic influences which inhabit all people's minds, and in particular, those who would exploit the public on what is apparently worthless trivia. A guilt complex perhaps, yet there is a psychotic necessity, just as it is with countless thousands of the so-called artistic or the entertaining element of this country—those who claim to be singers, movie stars, sex symbols, comedians, masters of ceremonies; yes, even beauty queens, all those who must, as a means of satisfying the psychotic desires of their mind, be constantly before the public, have the public praise, aplomb and applause raining upon them, to hear the clapping thunder of countless hands smacking together as a token of appreciation of whatever it is in the emotional content of this exhibitionism which has been deployed to these people. And these exhibitionists, these movie queens, movie stars, "the great ones" of the entertainment field, those who constantly adorn our television screen, become extremely sick and neurotic if they cannot constantly exploit themselves before the public.

Perhaps the purveyor of metaphysics, mental sciences, spiritualism and other particular religious expressions, yes, even those from the papacy on down in the great Vatican of Rome, have found the necessity of this expression within the neurotic or even the psychotic subterranean labyrinth of their subconscious mind. The insecurities, doubts, the frustrations and all of the other ego deflations which they have incurred during their lifetime and in their many lifetimes, have capitulated them into this psychotic dream world. They must constantly resubstantiate their own existence, their own life in the subterfuge of common public acclaim, to the subterfuge of exploitation into that

189

nether twilight zone which divides and differentiates logic and reason and an understanding of the common purposes and the psychology of life. Yes, and it even vilifies, deflates and otherwise expurgates and renders sterile any worth of any type of metaphysical expression, any type of mind science which might be contained in the involvement of any concentrated effort—be it mental, physical or otherwise—for a certain type of mental and physical concentration is necessary in any expressed will or desire.

Certainly not, however, do we resort to the subterfuge (and a common subterfuge) which is indulged in by all those who succumb into that false realm and dimension of the metaphysical metamorphosis where you become bug-eyed and repeat repetitious sayings all day long, or that you sit in the corner with your shoe toes pointed against your wall in this repeating, until you think you have a demonstration; or into any other expletives or other forms of these different metaphysical elements. This is indeed a dream world, the nether twilight zone, the realm of the neuropsychotic, to the person who is a candidate for the mental institution and the asylum, to the person who may be subjected to the different schizophrenias or paranoias which are a part of the Freudian psychology and the psychology of this day and time.

Even life itself, as I have stated, is an illusion; it is, to many people, a complacent pre-existence or it is an active existence, materialistic or otherwise, yet it is their world and they are completely engulfed in it; they are satisfied with it. In whatever fears they have of the future or even to death itself, they constantly and increasingly drown in the common amenities of their daily life. They may indulge in mild or heavy alcoholism, cigarette smoking, sexual promiscuities or whatever else they may unconsciously or consciously devise to placate the subconscious fears and insecurities and develop their own escape mechanisms, such as they are.

The reality of life as it ever presents itself, can be only thoroughly analyzed in an objective mind, and objective minds among the people of the planet Earth are hard to find—and even harder to come by. Such as it is, however, this is the premise of life—a premise in itself which demands an ever-increasing selectivity, an ever-increasing knowledge as to the mechanics of what life is; not just as it appears upon the superficial surfaces of the reactionary expressions of everyday life but rather, we must look into the causes, the true causes which always spring from and which are always powered and engendered by energy transpositions from an interdimensional cosmos, for this is a way in itself in which the Infinite Mind expresses itself infinitely in all forms and in all life.

Such it is for every human, that he must develop this selectivity in these numerous interpolations of expressive life forms as they materialize into his own life. They cannot be magical nor can he belie or otherwise deviate and change values and balances to suit his own need; but rather, that he should enlarge the circumference of his mental prospectus to fully incorporate in his understanding, in his learning and in all that he is in his evolutionary consensus, his position in the evolutionary consensus.

He must evaluate, use, or otherwise reject such interpolations of life as they are basic equivalents to a constructive evolution. In other words, selectivity is of the utmost importance; it is of paramount value to be able to constructively differentiate between that which may be negative, spurious or otherwise deleterious and possibly destructive. This is the morality of a progressive evolution. We are in the future what we are today at this very moment. No amount of shadowy evasions or subterfuges will suffice to make this future bright and illuminated, a wonderful place in which we can find a more free atmosphere of self-expression, a constructive atmosphere of expression, unless we become increasingly selective, unless we become increasingly knowledgeable of all of these factors which are involved and which I have constantly stressed.

No, I am not misanthropic. A misanthropist is a person who hates mankind and a hatred which is built up on all of the malfunctions of his life which has converted his subconscious into a labyrinth wherein there consorts the demons of hell, a person who is at any moment liable to commit some atrocious crime upon his fellowman, or a person who is to be victimized by subastral forces. Yet, let this subconscious mind be a place in which the perspectives of our everyday life can be played upon the screen of our intellect and we can become selective to the degree where we attain the true morality of our existence, that we can, in this time and in this moment, this present, begin our constructive future—a future which will be brought with the promise of our expressionary elements which relate us to their fullest extent, to their fullest deployment in all the constructivisms which are necessary to make our life palatable, to make our life justified as to the forward evolutionary movement of the Infinite.

For such as it is, there is an inviolate law of the Infinite constantly expressed in all things—the positive and the negative, the cyclic motion which is contained from the fourth to the third dimension and interplayed —or the cyclic motion, interdimensionally speaking. This positive and negative cycle is always manifested and it is one of the determinants which mandates our life upon planet Earth that every person must learn, must be able to differentiate which is the positive and which is the negative. This is the great morality of his existence. He will never be any more or less than his ability to be selective, not on the basis of certain selfish

192

conquests or that he is re-inflating a deflated ego but rather, his selection must be a constructive selectivity. It must be based upon fundamental knowledge of the cyclic laws of the Infinite which express themselves in two dimensions simultaneously and in two directions or they express themselves infinitely into all directions without time or space. Such is cyclic motion. No, it is not seemingly a movement which is without form or motion for in that world all is comprehensive, all is tangible. There are no intangibles, there are no fears, unless these are fears which are self-instigated through lack of knowledge, through lack of proper purveyance of your own evolutionary movement to whatever it is that you have laid down and formulated as to the constructivisms or the negativisms of your life.

No, these are not threats; they are not coercions. They are simple statements of fact. That I love mankind in all its forms is without question or I would not be here; I wouldn't be talking as I am. I would not be devoting the last moments of a temporal earth life into completely justifying the cause of this Mission. To you and to all peoples of this Earth, whether they are large or small, and whatever sex or age they may be, whether they understand or whether they would reject all that I have to give to them, yet within the circumference of my Mission I am able to express to them in certain ways and in certain factors a positive plus into their lives—a spark, if you please, which will enable them in the future to take the proper pathway which will be the bias which will show them the right course to take. It will enable certain agencies from Higher Worlds to contact them and to place them into more constructive positions—to give them some insight into the mechanics of interdimensional law.

Yes, indeed, all this is a part of all which I have come to this Earth at this time to express. Whether or not you can accept these things, whether or not you can believe, as I said once before, "If you believe in Me and believe that which I teach then you will have immortal life," will make the difference. It was true then; it is true today, for nowhere else within the circumference of an earth world can you find that which is most necessary for you to help you evolve from this Earth except that it be given to you from those who have mastered this world and gone beyond the dimension of its circumference; they who have mastered the interdimensional science and have become part of the Infinite Cosmos.

These are the Master Minds which you have epitomized in your religions, in your earth world documentaries such as they may relate to the supernatural; yet, they are people; they are in a sense human beings. No, they do not possess bodies as you do; theirs is not a composition of the common atomic forms related to your ninety-two natural elements of this earth but rather, theirs is a body made into configurations of more infinite substance—energy forms which are beyond your comprehension.

Until that time and day when you do develop that comprehension which will likewise develop the necessary body, the mandatory vehicle which will enable you as a conscious entity to function in a Higher World, never cease in your seeking—this is my message. You may take it or you may leave it but if you have listened, I know you will never leave it.

Inborn within any living form as it is in the world today is that positive bias. So it has been throughout the reaches of time into other planetary systems and to other worlds—the bias of infinite progression is always incumbent and is part of even the smallest cell, the smallest form of life, a bias which will, in its natural and subsequent development through numerous lifetimes, adapt to numerous forms or even incorporate as embodiments in other forms and will begin to expand in its circumference will begin to include factors and intellectual perspectives in a progressive evolution.

Were this not so, the Infinite would cease to live; it would cease to be expressive as it is in all forms and in all life. It would cease to be the power behind the appearance through the normal function of all, not only of third-dimensional forms but forms throughout the interdimensional cosmos. Were this not so, there would be nothing, there would indeed be that void spoken of in Genesis—a void in which even the godform as it has been so deified could not possibly exist in the nothingness which was presupposed. How baseless, yet the beginning or the end is not in sight.

Even those who have aspired into a higher way of life do not know the ultimate answer or the beginning, yet a beginning and some answer there must be. Life in any form must be justified and as to the personal quotient of each man's mind. This justification must be a constant compromise as to what he considers virtue and what he considers evil, or that his selectivity, his mind function enables him to transpire and transpose evil and good as he sees fit. So he has begun his greatest of all immoralities—that he may create evil for his own purpose—and evil it is when he centralizes his existence into personal gratifications such as are expressed by (so-called) mind sciences.

The acquisition of wealth (as taught by these false teachers) is a complete centralization of desire, a complete centralization and a revolution within the citadel of his own mind wherein he completely surrounds himself with a fortress, an impregnable forest of thorns which cannot be penetrated by logic or reason. He is completely immersed in self-gratification, the material desires. He has thrown selectivity to the winds and he completely absconds into the dimension of emotional gratification. That is the promise of metaphysics—metaphysics which teaches the acquisition of such affluences and amenities of life which gratify personal desires. It is the lust of life and it is the rape of spiritual consciousness.

So do not hesitate. The greatest Master in the Infinite Cosmos is a Master who is first master of himself who has learned to conquer what is the most obvious, the most dominant characteristic of his nature, the beginnings of his nature, the survival of the fittest; for within the tenets of this survival principle are impounded and incorporated the most primitive desires, the most lustful desires, the lusts of self-acquisition of wealth.

To exploit these common human weaknesses, to attract those to you upon the promise of fulfillment or that you can attain whatever it is that you desire on the basis that this is a personal acquisition, a personal triumph, this is indeed a horrendous crime.

Such practitioners or exponents of so-called metaphysics or sciences should be outlawed and branded as common criminals. There should be strong laws to curtail their activities for they are indeed purveying a common drug—a drug more vicious than heroin or any other of the drugs used by an addict. It is an opiate, it is a drug so powerful it will destroy any human who enters into its usage; he will become addicted to it. Yes, he may even live in a palatial mansion on a hill and he may have a butler and a chauffeur and many cars and may dress in the height of fashion and portray all of the affluences of his society, yet this is indeed the greatest of all addictions—an addiction born out of the common subterfuge of survival—that we survive or that our ego must be placated to that point where we triumph over the common insecurities of our everyday life.

How well it would be if we would triumph over the mastery of ourselves to the point of self-admission, that we must objectively face ourselves in this self-admission through inception, through admission, through tacit surrender of whatever it is we absconded into the labyrinth of the material world.

These are our sins, our karma and our purposes. Our necessities to extract ourselves from this labyrinth are well-defined, but first there can be no compromise over the mastery of self and its primitive desires, for the subconscious, as it was expressed in any lifetime or in the present is still primitive, is still biased by the oldest of all desires—the desire to live, the desire to procreate, the desire to obtain supremacy, to relieve the psychic pressures of death, of other fears and insecurities of the daily life.

The Aquarian Age

During the twentieth century and especially in the latter half of the century, there has been a tremendously accelerated pace in all factors and aspects of mankind on the planet Earth—a great renaissance, as it were, a tremendous abundance and redundancy. Along with electromechanical discoveries and developments, the Space Age has blossomed into full flower and along with the more progressive and intellectual developments there has also been a great retrogression —all of which again amply fulfills certain Biblical prophecies. Also as part of this tremendous expansion, there has been a great resurgence in the mystical elements of man's life—elements which do not more properly belong to orthodox religious structures but should be classified as a resurgence of mystical beliefs coming from the more ancient origins of mankind.

Heading the list of these more popular mystical expressions is astrology; in fact, astrology does proclaim this particular century as the beginning of the Aquarian Age, one of those two-thousand year segments of a twenty-five thousand year cycle of our solar recessional. In any city of the world, in clothing shops, boutiques, offshoot cultisms and subcultures, manners of attire, wearing of long hair and beards, daily habits and expressions, drug usage among the younger people of this generation are all inclined toward the mystical side of this Aquarian Age—and again more properly understood in a broad consensus of cyclic movement and interplay within the interdimensional cosmos.

Today almost all peoples and particularly the younger generation, have been caught up in this mystical resurgence and there are many neophytes, as well as "professionals"—all coming forth with numerous expressions and interpretations as part of their beliefs in this Aquarian Age. Aside from astrology, there has been a resurgency in witchcraft and its attendant cult-isims of spiritualistic mediumship, water dowsing, locating hidden wealth and minerals by psychometry, paranormal activities and investigations, etc. It has been estimated there are one million "witches" in the United States.

One life factor, however, which still remains domi-nant and seems to have suffered little change is death —man's natural fear of it and a great desire to alleviate or mitigate this fear by whatever means or methods which are existent. Strangely enough, it might seem, the orthodox religions of the world have not supplied the answers to either the mystical resurgence or to the enigma of death and other than a supposed heavenly abode in a fanciful city of New Jerusalem, no other reference has been made by these religions to the pos-sibility of life after death, its consistency, et cetera.

There has, however, appeared numerous articles published in newspapers and other communicative media, books, et cetera, and personal experiences of these people who have had actual living experiences in the next world. Aside from spiritualistic mediumship, most of these are individual experiences which can to some extent be verified. One of the most commonly expressed mystical or psychic experiences experienced by many people is astral flight, the apparent traveling from the body in the psychic or astral body to some distant point or place, then returning to the body and recounting and describing the experience, et cetera.

One of the more widely known cases relates to a young man who, while medically trained, became an inductee in the Unites States Army and during one of the current viral epidemics, he contracted in 1943 a severe pulmonary infection and actually died from this condition. Then for five days and, according to several

199

doctors, the body was apparently lifeless. It was cold, still, no pulse or respiration. All signs of life were gone, yet five days later the young man did return to his body, reactivated it and for many years lived a full and useful life.

Most startling however was his description of what he did and what happened during these five days and this narration particularly verifies the descriptions of Celestial Beings and the great universities I have described in the Pulse of Creation books. Briefly, he tells at the moment of death, a beautiful spiritual Being such as I have described, entered his room, took him from his body, then, in astral flight, over towns and cities to a Higher World where he saw many spiritual people going to great universities, being taught in all the arts and sciences. He even saw the atomic submarine, "The Nautilus", being built ten years before it was constructed on Earth.

Also, later on after returning to his physical body, in company with a young man, he went to a city he had never been in before and correctly identified a certain white painted restaurant, its location, et cetera which he had seen while in astral flight with the Being. The full account of this story may be had in the December, 1970, issue of "Fate" magazine.

Now all this astral flight and the proposition of this flight may, according to certain statements I have made, seem to be contradictory, so some clarification is in order. I stated that in the astral or fourth-dimensional world no one travels anywhere, which is quite true. The interdimensional cosmos is composed of many dimensions or layers—each dimension can then be further subdivided. Then each subdivision is further divisible and so on, ad infinitum. So far as this interdimensional cosmos is concerned, most of it is composed, shall I say, of energy in different kinds of cyclic forms—all harmonically interrelated.

In the third dimension, one of the plane surfaces of the interdimensional cosmos, these cyclic wave forms assume the subatomic form or atoms, each atom in its specific gravity expressing a facsimile of the spectrum from whence it sprang.

The life experience of any person living on an earth world does, abstractly speaking, assume an interplay of atomic forms which have become familiar to him in his evolution such as houses, trees, mountains, other people, et cetera, and his daily life consists of moving around with a limited number of atomic forms—all of which are immediately related to his consciousness as oscillating wave forms which again are oscillating in a limited spectrum relative to his third-dimensional world. Therefore, in any manner or way that he leaves his physical body and enters the spirit world or the fourth dimension, he will meet and live in an infinitely filled dimension of oscillating cyclic wave forms which he cannot directly relate as consciousness. They mean nothing to him.

It must also be remembered that in this fourth-dimensional cosmos there is no time or space. Because of the previously described interrelated harmonic oscillations taking place in the net sum and total of the interdimensional cosmos, everything and anything exists simultaneously anywhere; that Consciousness can integrate certain wave forms pertinent and relevant to, and as the substance of whatever is being objectively conceived in consciousness.

Therefore, any Being who can interpret this interdimensional cosmos does not go anywhere. He merely tunes Consciousness to whatever he wishes and it's there; he's in it, oscillating with it, just like tuning your radio or television. However, an earthman unfamiliar with the interdimensional cosmos (the spirit world) cannot interpret or "tune in" to this oscillation in a way which would be a tangible form of consciousness. He must therefore, by his own efforts or lack of them, live in a dream world of memory oscillations contained in his psychic anatomy—or a Being such as a Unariun Brother can come to his aid and by means of projection, actually project into his consciousness certain energy wave forms which are relative to his (the earthman's) understanding and his past life experiences. Therefore, he does consciously seem to travel. He may even seem to see cities and towns all relative to his earth life, all of which are actually auric or electromagnetic radiations of these material objects related to him and rectified in his consciousness—a direct facsimile analogous to television transmissions projected into your home and interpreted by your television set as forms relative to your consciousness.

So you see, here again correct and proper analysis is most necessary when we enter into any of the current aspects of human life, whether they are everyday, mundane, physical life experiences or whether they relate to some of the more mystical aspects. Again, all such knowledge which can correctly solve these enigmas and riddles will be knowledge most pertinent and valuable to you when you leave your physical body, either temporarily or permanently. And that, dear friend, is the purpose of Unarius—to give you this knowledge.

I might even add: if you don't begin to develop this knowledge, to use it, to make it part of your life, then you're going to keep on revolving around in this little material hell-hole until you disintegrate, sucked dry by the subastral forces, etc. This is not a threat; it's a prediction and a very accurate one, based upon the life experiences of countless millions of people.

We the Unariuns are here to aid you in all possible ways, and yes, sometimes the impossible; but primarily, we depend upon your efforts, your dedication.

To Ioshanna —

Yea, even as ye hath done it unto these
so that it shall be done unto thee
And that thy sheep shall be fed with the
pure waters of Spirit
Which flows from the headwaters —
uncontaminated by the lusts
of the earth world.

Fear not that ye lack the wisdom
to relate to thy brethren
For thy storehouse hast been well-filled
with the full measure
of Light and understanding.

Many are they that await the word
that issues forth from thy mouth
For with the Word cometh the Light,
the Power and Radiance
From God's great Infinite Mind.

Rejoice ye that the day of the awakening
is at hand.
Fear not for I am with ye always —
even beyond the end of time !

12/'72   *— Raphiel*

In Retrospection

During the many years in which I and my wife Ruth have been involved in the administration of the Unariun Mission, there have always remained several bones of contention which seem to affect rather adversely, the proper understanding of the Unariun Mission, and we have, from day to day, throughout these years, constantly been confronted by questions or emotional attitudes of those who have not only just begun to study Unarius but in some cases students of rather long standing.

Misconceptions are anticipated and seem to be at least temporarily justified on the premise that Unarius is an interdimensional science which totally involves the human evolution in all aspects from the most primitive beginnings to a more ultimate configuration, wherein any human could advance into a Higher World far beyond the boundaries of the third-dimensional world, and at this stage, any human is particularly vulnerable to numerous emotional temptations which could and occasionally do, temporarily at least, lure him back into the mire of the past.

As all such first steps, therefore, are pivoted upon any existing (or former) religious beliefs, it is most necessarily logical to begin the advancement of an interdimensional comprehension upon the basis of such existing beliefs, bridging, so to speak, the present with the beyond. Religious beliefs, however, do not exist in any such Higher Worlds; neither do the common reactionary means of existence as they are so compounded in a third-dimensional material life. Unarius, therefore, means not only a complete separation and abandonment of all such earth life attitudes, but actually means the complete rebuilding of any such individual human into a pure energy or spiritual form which can live and be totally integrated in a Higher World dimension.

While the purposes of such an evolution and re-building have been most adequately served in the Unariun liturgies, there still seems to be a lack of a proper understanding or approach to the problem of logically beginning the rebuilding process. The first series of books, "The Pulse of Creation", most ade-quately described how certain numerous human indi-viduals ascended, so to speak, into one or more of these higher in-between Spiritual Worlds. Here the more familiar solid mass earth world configurations were recreated from the pure energy which infinitely fills the timeless, spaceless dimensions of these Higher Worlds. The proposition as to whether these worlds or cities actually exist is purely academic; also these cere-monies, observances and such, are likewise tailored in a symbolic form which can suitably represent all that which has transpired in former earth lives.

In a more complete abstract evolution and beyond these seven Spiritual Worlds, such symbolic forms do not exist, for such symbology exists only as a necessity for a less highly-developed intellect, wherein all such symbologies, as they currently exist in earth worlds, are most necessary for the common concourse of life, and on the basis of wave-form oscillations, make it possible for any human to exist in such an earth world, whereas in more advanced worlds such sym-bologies are not necessary and do not exist.

Consciousness is a direct attunement with Infinity or at least, Infinity in its proper spectral form with that particular relative dimension. As you can now see, the proposition of an interdimensional life in a Higher World is indeed a far cry from the emotional-reactive life lived on a lower earth plane, and any such evolu-tion for any human into this Higher World would be extremely difficult if not impossible, and without help, comparatively very few would ever make it, even in extremely prolonged periods of time—say hundreds of millions of years.

For purposes, therefore, to best serve third-dimensional humanity and to serve Infinity, to help promulgate logical human existence in an infinite number of ways, there was brought about throughout the countless billions of ages the formations of great groups of Advanced Beings. Beings do not live in Higher Worlds in idleness. Theirs is a constructive way of life, very highly evolved, technical and scientific, wherein they deploy their power and knowledge in the best evolutionary purposes.

Most vitally interested in this earth world evolution and with the concourse of humanity which has and is reincarnating and evolving through its third-dimensional portals, is the Unariun Brotherhood—actually many millions of these more highly-developed spiritual Beings formed together as an organized group deploying all manner and means at their disposal and as a development of their present stage of evolution. Now, none of these Beings could or would reincarnate physically as a personality into the earth world, nor would it be logical for them to do so. The sheer intelligent energy-power of their total embodiment would always seemingly transform or transmute the human form or the human life. Likewise, sanity and reason in any human who viewed this Super-Being would also be transmuted or transformed.

Therefore, a transitory vehicle must be used to deploy and convey the necessary information and power to any human who has most advantageously been preconditioned in many manners, ways and forms to the infusion of this intelligent power into the prospectus of his life. This vehicle must then by necessity in all appearances be an ordinary human of the same form, wearing similar clothing and adopting all other habits and appearances of that particular time; however, there are differences. Through the higher psychic anatomy, a kind of channel has been formed wherein the Unariun Brotherhood project into the conscious mind of this person the most necessary power and information which is again by various means relayed into the material world.

No doubt you have watched television. Now you know that the people and other things you see are not inside the television set; they come to the set from a broadcasting station by means of electrical waves which carry certain other electrical waves which, when properly sorted, integrated and amplified by the electrical circuitry within the set, appear on the screen as pictures—the sound coming from a nearby speaker. The same proposition in principle is analogous to the way the Unariun Brothers work with the people of the Earth.

By means of mind projections, certain types of electronic apparatus, and such, they can and do, in a thousand different ways, project certain electronic impulses into the psychic anatomy, the subconscious and the conscious mind of countless millions of people. Some people call the realizations of these impulses intuition or hunches, and in more isolated instances, certain people have claimed to see God or Jesus when the picture of one of the Unariun Brothers was projected onto their inner "mind screen".

Throughout many ages of man's written history, several of these especially prepared and constructed humans have appeared and delivered whatever particular Mission was theirs to do, a particular Mission tailored for the people and the time. One of these more recent persons or emissaries was Amenhotep IV (1300 B.C.), then Anaxagoras of Greece (500 B.C.), and more finally, Jesus of Nazareth, this man, Jesus, becoming the cornerstone in all orthodox Christian religions.

Actually the man Jesus, as He was so portrayed in the Bible, did not really exist. He was in every way and in every sense of the word, directly opposed to all existing religious beliefs, either then or to such subsequent developments which have led to the present Christian orthodoxy.

The appearance of the "The True Life of Jesus" book, in 1968, was therefore more than a fortuitous circumstance. The entire Unariun Mission, the compounding and publishing of its liturgies was brought about by the fourth special person or Emissary. Again a psychic anatomy was properly reconstructed from portions of previous Emissaries, or reconstituted facsimiles in energy form and with a suitable channel. This fourth Emissary was most appropriately born in the beginning of this twentieth century wherein, during his life He would, at the appropriate time, enter into and consummate a certain number of specific purposes. This fourth Emissary, known as the Moderator of the Unariun Mission, also bears an ordinary earth world name, Ernest L. Norman. Also and very appropriately, through reincarnation and as a means of substantiating her reconstructive evolution, the person known as Ruth or the wife of the Moderator (formerly his beloved Mary) was also joined with him as a polarity, and in the numerous ways through the channel of her mind, she also brings power and intelligence into the Mission.

The entire context of the life of Jesus, who He was, His Mission, together with His beloved Mary, now Ruth, is most adequately served in the book, "The True Life of Jesus". Likewise, as of this time, those who were involved with Him and the crucifixion have, through reincarnation been gathered together, each one individually knowing his or her place and, through scientific principle, are working out their negative indebtedness (their individual accounts being given in the sequel book, "The Story of The Little Red Box" and in fuller accounts in the "By Their Fruits" books)—one of the primary purposes of this Mission. Yet, there are more ultimate purposes, one of which is to place at the disposal of countless thousands of persons, either in the world at this time or appearing in the future, all pertinent and relevant information and facts of inter-dimensional evolution—a true creative Infinity which should and will displace the common religious fanaticisms now currently existing.

Therefore, to all those who approach and touch Unarius—do so with an open mind; try to lay aside, temporarily at least, preconceived notions formed from out the primitive past. Infinite Creation is always moving and at incomprehensible speeds, always forward or positively biased. It always remains an individual prerogative to always be cognizant of this fact; that which is of today must pass before the morrow begins. Life is a series of reincarnations which, when these lives are sifted, that which is good can form the steps which will lead any human into a higher evolutionary stage.

Seek, therefore, and ye shall find. What perplexes today will be understood in the future and again, in sequential evolution, becomes the past.

Addendum: It is almost totally incomprehensible that as of today most of the 800-odd million people who call themselves Christians are surrounded by, live in, use, or in some way understand the existence of a vast and scientific civilization, and with a certain sophistication with this scientific civilization, still believe in, venerate, worship, hold sacred, in toto, a certain book called the Bible which has been compounded from old legends and fables wherein certain ancient peoples in rather primitive beginnings compounded their religious beliefs into the construction of god-ideosophies or god or gods. Herein, this god not only was believed to have infinitely great magical and creative powers, but strangely enough also exhibited a thousandfold in intensity all depraved human emotions of lust, murder, fornication and an overwhelming vituperous anger, and who used coercion, trickery, lying, subterfuges, etc., to gain his ends within the very limited circumference of the Israeli nation, and which also affected other surrounding nations in the shambles of the godlike intemperances and indispositions.

Furthermore, and even worse in this biblical portrayal, this same vindictive, vituperous god purportedly sent his son into this nation for some rather obscure reason or reasons, not entirely clarified or justified by the very apparent opposing doctrines and philosophies represented by this supposed son. We will overlook a common form of adultery used by this god to enable his supposed son to make his ingress into the world.

However, it is difficult to overlook the portrayed incidents contained in this biblical version of the man's life, a charade of miraculous healings done senselessly with no apparent purpose other than some sympathetic compassion and very much against a sound psychological understanding which would also involve knowledge of reincarnation, evolution and all other human factors. Also while this man was sufficiently powerful to perform such miracles, he apparently lacked or didn't use the knowledge which would, in either case, be most necessary before any miracle could be performed. This total fanaticism is now exhibited and believed in by the total Christian community!

Yet, this is not all. As a ghastly climax to this man's life he was, through mob violence, nailed to a wooden cross, then later the following Christian hierarchy contrived a ghoulish nightmare from this murderous incident, that this man's blood had now automatically washed away the sins of not only those who had or did now believe, but to the hundreds of millions of unborn Christians who would later embrace this fiendish contrivance.

And so as of today and this hour there exists the most incomprehensible enigma ever written in the annals of human history of hundreds of millions of humans who are surrounded by and whose lives are lived in a complete valedictory of scientific creation, yet have wholeheartedly absconded into the hellish nightmare of lies, lusts, derelictions, fornication and murder, compounded and portrayed in the fabrication of historical legend they call their Bible—a fanatical religion steeped in the blood of human sacrifice and raised upon the altars of their churches as the highest epitome of their social structure, their criteria as to the moral and intrinsic worth of any human who does or does not believe!

Psychokinetics

It has often been said and quoted that the average individual does not use even a small fraction of the more ultimate potential of his brain capacity—roughly five or ten percent. This, in one respect, is quite true; however, in modern science, and particularly in a newer dimension of science which relates to the exploration of various nerve impulses which are connected through the brain and nervous system of the human anatomy, there is still inconclusive proof, in fact, a serious doubt that the brain is capable of regenerating an electrical impulse. Therefore, it can be concluded that power, as a potential of energy, so far as the brain is concerned—and in particular to its ability to generate a potential—is figuratively nonexistent. The truth of the matter is, and as postulated in the various teachings of Unarius, this brain power comes from the psychic anatomy, from a direct attunement with the sum and total of consciousness as it is expressed in the oscillating wave forms which form all and various sundry expressionary elements as they are so contained in the psychic anatomy. Therefore, the potential of energy, as it may arise, or the capacity to project energy into any particular dimension or dispensation resides entirely upon the intelligence quotient, as it were, which resides within the psychic anatomy itself, and not within the brain.

Perhaps the best way to illustrate this proposition is in a personal experience. Before relating this personal experience I must make a certain qualification; that inasmuch as to justify what I am attempting to explain to you at this time, I am, in this personal interjection, not trying to boast or brag but to point out to you the development of consciousness as an expressionary element in explaining what is meant by the fundamental principle of psychokinesis, or a potential of energy in a projected form of consciousness into any particular expressionary or third-dimensional adjunctive which demands or necessitates the expurgation of such extra-terrestrial energies. The beginning of this incident or personal experience was back in the 1930's during the depression days, and in those extreme times as a man with a family, I was forced by necessity to seek out various and different jobs whenever and wherever they so presented themselves and in whatever menial capacity, regardless of status quo. It was in this respect that I found myself more or less as a common laborer, working upon various construction efforts which entail-ed the building of a large house for a multi-millionaire airplane manufacturer. There was, in the course of this employment, an association with several different workmen who were so similarly involved in this job, and I struck up an acquaintance with a very fine chap by the name of Frank Porter. He and I became rather warm friends.

Now one of the particular implements which I used in my various tasks was a certain shovel. This shovel became sort of a pet thing with me, in a sense of the word, because I developed a certain feeling or touch with it. Frank also must have found a similar rapport with this shovel for he very often purloined it from me whenever he saw that I was not looking, and the shovel was conveniently located so that he could acquire it even temporarily. The constant succession of these incidents somewhat aroused a little irritation, and one morning when I arrived on the job and began taking up the daily tasks, I found my shovel was missing. And at that particular moment, something happened inside— inside my mind, so to speak. I had a strange and very powerful and, for the moment, a rather awe-inspiring feeling or a connection which I had not, up to this point in my life, had in that similar manner and fashion; neither in that strength.

I sought out the place where Frank was working nearby in an excavation, which was destined to be a fish pond. I walked over to the excavation and as I approached, Frank turned around; as he turned, I said to him, "Give me that shovel." As he looked at me, a look of terror swept over his face, his jaw sagged and his eyes protruded. He automatically handed me the shovel, and as he did so, he started to scramble out of the excavation. After he had climbed out, he ran pellmell back to the corner of the house nearby. Then for three days Frank avoided me; all I saw of him were his heels disappearing around the corner.

Finally, during lunch of the third day, I cornered him and asked point blank why he avoided me. He answered me this way, and I will quote his words: "My God, man, you should have seen your face when you looked at me and asked me for the shovel!" He said, and I quote again, "I was never so terrified in my life," and I could well believe Frank! Then it took some time to overcome the last vestiges of this terror in the subsequent days which followed. This particular incident left an inward impression on me which I never forgot.

During the war years, I learned a much fuller meaning of this feeling which I had within me on that particular day. No, it was not as I thought it was at first— anger; it was something other than anger, because the morning when it first occurred was a morning in which I was in some sort of a trance-like state. I had not fully awakened from my night's sleep and was still partially hypcognic, and in that moment of orientation, with the shovel and the various incidents concerning it, I had somehow managed to acquire a connection which I had not previously acquired. During these terrible years of the war where I went in and out of the various dance halls, churches and other places trying to work up and orient the inward feeling or the creative urge for spiritual expression and teaching, I had numerous occasions to contact this inner feeling and this inner source of power, and whenever it was done and done correctly, the results were immediate and sometimes awe-inspiring.

I remember one occasion in particular when I walked into a room where some sixty people were busily engaged in ordinary conversation as they sat around the room. As I braced myself before entering, the same feeling I had so many years before manifested itself and when I walked into the room, the buzz of conversation stopped and immediately every head was turned toward me. They looked at me and they could not comprehend; somehow there was something there they did not recognize. Afterwards, towards the close of this evening, I had the opportunity of reaching each person individually in a personal message which involved a great amount of demonstration in the various paranormal fields of clairvoyance.

It was very much this way when I entered a church and ascended to the rostrum. Being naturally introverted in some respects, due to some childhood repressions, I frequently and almost always called within myself for this secret contact of power, and when I felt this power come in, my being was completely transformed and I walked upon the platform, always, in this moment of transcendency, able to give a completely satisfactory lecture, or the various reading services entailed in these different churches always culminated very satisfactorily.

This same expression held true in so many other different contacts and ways which were most necessarily made manifest in my daily life. As a rather shy and retiring person, it was most necessary that I call upon this extra inward fortification to bolster up what might otherwise have been a very introverted expression of life. It was in this way that I did succeed, in these moments at least, in contacting the inner power of the Inner Self, the power which is only vaguely dreamed about by various psychological or physiological differentiations, and which is never really ultimately achieved except by people who understand the true principles of life—the regenerative principles which we have taken great pains to explain to you throughout the various lessons and books of Unarius. After a while when this wisdom has permeated into your consciousness and has re-formed, so to speak, or helped re-form the various portions of your psychic anatomy, you also will be working in accordance with these principles; then indeed, you too will find that inward contact, just as I have, and you will realize the tremendous potential of human consciousness which can be so constructively expressed in that particular moment.

I might further add and enlarge many other instances wherein people have described this power of psychokinetical projection which they have felt when coming in contact with me. Some of them described it that they tingled all over or that the hair on the back of their neck rose straight up. Other people, too, felt it in different ways. As I analyzed each and every one of these instances and incidents, as I went from one person to another or from one group to another, it seemed that each contact ultimately served a most intelligent purpose. There were also other numerous and automatic functions in this contact of the inner self which were manifest in various physical reactions with different people. They would, on contact with me, break out in a very violent perspiration, their clothing would become wet with this extreme condition. They would pant for breath, their faces and countenances would be flushed to a crimson tint with the excess blood which was rushing here and there about their body. This was a healing process which cleansed and purged the physical aura and helped tremendously to purge the physical body from various different vibrations of static energy which were most necessarily incurred and so incumbent in the physical anatomy of the materialist. Furthermore, in almost all instances, this particular expression of power could be turned on and shut off at will. I demonstrated this on many occasions. In some instances I was classified as a devil for expressing this particular expression of healing power. Now, of course, I realize that whatever this particular ability is, it must also consist of a tremendous amount of preconditioning and usage, particularly in former lifetimes or in Spiritual Worlds which must, by necessity, demand and use such extensions of mind power. It is not necessarily a relevant expression which is found in the material dimensions, for the material man does live largely off the various emotional expressions as they are found in the usual mandates of the physical world. He seldom, if ever, does extend the power of his mind beyond the immediate demand of whatever is presently so confronting him at the moment. He therefore is largely ignorant of any particular spiritual dimensions and the power within one's self to achieve a certain function with these higher planes of expression.

The psychokinetical power of projection which seems to stab like a ray or a beam from the person who has learned to acquire and use this power—this extension of power as a ray or a beam, does accomplish the immediate objective for which it was so contrived. Furthermore, this expression is, as it could be wrongfully assumed, neither destructive in its potential nor is it ever destructive in its results, but always achieves a more ultimate and sublime accomplishment, inasmuch as it can correct, heal, adjust and otherwise remove various mental and physical aberrations from those with whom it comes in contact. In other words, it is a general summation of consciousness projected in such a ray or beam which extends from the mind of the person who has so learned of these things and learned of their usages. This is the prime purpose of Unarius; to first separate you from the various emotional ways of life lived by mankind; separate you from the symbologies, the thought patterns which mandated your existence from the past evolutionary lifetimes which you have formerly lived, up to the point where it can be reasonably assumed that you have sufficient intelligence to be more or less in control, shall I say, of the various and numerous essences or abstractions of thought consciousness. Until then these emotional experiences are most necessary to give you the necessary libido or drive to constantly reconstitute and reorient yourself into your daily world.

However, after the accomplishment of the more infinite prospectus, these various emotional devices are no longer necessary or needed. In fact, if they are maintained, they can be quite destructive, particularly if you have somehow envisioned within your mind, in the Higher Worlds or in moments of spiritual consciousness, this higher way of life, the reconstructive power of mind extension as it oscillates infinitely; it would be much better if you had never conceived these things or approached the threshold of conception of these Higher Worlds. For such is the way, in direct ratio and proportion as to whether you do recognize the existence of these things or whether you are actually accomplishing some practical purpose in the utilization of these things.

Any time that you renege and go back to the more emotional vicissitudes of your former earth life, these will then indeed be, conversely, so much more destructive than they could ever have formerly been. The power of psychokinesis or mind projection is very real and it is very tangible. It can be expressed and utilized or usefully used by people who have learned of the principle and function of energy in its higher inter-dimensional states of expression. It can be a generally expressed form of intelligence which can reach thousands or even millions of persons simultaneously or it can be in a more specifically directed area of contact to one singular person, but whichever way are the manner and form in which this psychokinetical projection is so utilized, it should always be constructively projected.

Should this projection ever proceed or be utilized in such personal expediencies which can relate the individual in a personal way to these expressive forms, then he is indeed practicing, either directly or inadvertently, certain pagan or barbaric practices which reside in historical contents as witchcrafts or sorcery, and the results of this expressionary form of psychokinesis is most surely leading this individual on the downward path to oblivion.

It might be well at this time and serve a useful purpose to explain one of the paradoxical recounts in the life of the Nazarene which was lived some two thousand years ago. This is the story which directly connects Jesus in the act of flagellating or whipping the various mendicants who were selling the votive offerings in the courtyard of the temple in Jerusalem. The scene so depicted, relates that Jesus used a whip to chase these various merchants from out the temple. Now it is inconceivable that a man of the mental stature of this Nazarene could ever resort to the expediency of using a whip on his fellowman. It is here that we can express perhaps the greatest example of the use of psychokinetical projection or mind power. Just as I had formerly done in that first episode that connected me with the shovel and Frank, so did Jesus, when He walked into the temple courtyard on that particular moment, become so concerned with the various negative energies which were so concurrently being expressed by these various merchants. The sum and total of all this negation, like a huge and overwhelming tide of thought and energy, began to overpower Him and He immediately resorted to the fundamental expression of his existence. He contacted the Higher Self, and in this moment of contact, the great inflow of psychic energy began to infiltrate, to permeate, to exude and to project itself through His personality as He walked through these various merchants. They, too, just as I had previously done when I walked into the room of people and they had turned and looked at me, so did these merchants turn around and look at Jesus, and what they saw struck terror in their hearts.

They could not explain the tremendous wave of positive power which swept over them. They only knew, just as did Frank, that it was something new; it was terrifying in its newness and in its differences. So they fled while Jesus stood among them without so much as raising his finger. This was the true way in which these merchants and mendicants were chased from the temple courtyard. Like Frank, they fled in terror from the unknown from the great power which swept over them and which was something they could not rationalize in the various reactionary moments and memory forms of consciousness in their daily earth lives. Then, in the natural course and event of history, people who read or heard this story recounted could not understand it in its true form; they had to symbolize it and in the course of history, the whip was placed in the hand of Jesus, whereas, He was, in his entirety and in his true self, expressing merely his high knowledge and his contact with the Infinite Mind of the great concept of Intelligent Creation.

This is typical of the way in which man degrades in his physical expression of life and institutes a common decadency in all forms of expression which relates him to the higher Spiritual Worlds of expression. The Bible is full of these numerous instances; in fact, religion itself is a decadency of these principles. Instead of instituting the correct knowledge, function and usage of personal contact as a form of consciousness with the Infinite Intelligence, the false world of symbology has been reinstituted. The mirage of personal intercession and various other devices deistic in nature, are substituted for the true meaning of life.

If you would obtain immortality, if you can conceive a better way of life, a way to live beyond the grave or worse, the grave of materialism as it expresses and is being expressed in various currently existent forms throughout the world today, then you must learn not only of Infinite Consciousness and the principle of creative expression but that this same similarity must also exist within the conscious form and way of your own life. Life in the Spiritual Worlds and in a more advanced state is not one of idleness nor spent in lassitude. It is a direct conjunction of constructive consciousness with Infinite Intelligence. It is proportionately equalized with the principle in which the personal individual so sustains himself in this higher form of evolution. Consciousness and the re-expression of consciousness is the mandate of his survival; not necessarily the carnal lusts, the primitive instincts and the various libidos which are sponsored by the emotional experiences from the material planes.

We can only constantly reiterate these principles, these truths and these wisdoms until they become infiltrated in consciousness—until they become constructive forms of consciousness within your psychic anatomy and you will constantly and eternally oscillate with them.

Other works by Ernest L. Norman:

The Voice of Venus
The Voice of Eros
The Voice of Hermes
The Voice of Orion
The Voice of Muse

The Infinite Concept of Cosmic Creation
Cosmic Continuum
Infinite Perspectus
Infinite Contact
Truth About Mars
The Elysium (Parables)
The Anthenium   "
Magnetic Tape Lectures
Tempus Procedium
Tempus Invictus
Tempus Interludium Vol. I

Also a publication, now reprinted by
Unarius Publishing Company:
The True Life of Jesus of Nazareth (1899)

(The Sequel): The Story of the Little Red
Box